Elderly Population in Modern Russia

Irina Grigoryeva • Lyudmila Vidiasova
Alexandra Dmitrieva • Olga Sergeyeva

Elderly Population in Modern Russia

Between work, education and health

 Springer

Irina Grigoryeva
St. Petersburg University
Saint Petersburg, Russia

ITMO University
Saint Petersburg, Russia

Alexandra Dmitrieva
Alliance for Public Health
Kyiv, Ukraine

Lyudmila Vidiasova
ITMO University
Saint Petersburg, Russia

Olga Sergeyeva
St. Petersburg University
Saint Petersburg, Russia

ISBN 978-3-319-96618-2   ISBN 978-3-319-96619-9   (eBook)
https://doi.org/10.1007/978-3-319-96619-9

Library of Congress Control Number: 2018957077

This Springer imprint is published by the registered company Springer Nature Switzerland AG
The registered company address is: Gewerbestrasse 11, 6330 Cham, Switzerland

Acknowledgments

The study was performed at ITMO University (Russia) with financial support by the grant from the Russian Science Foundation (project №14-18-03434): "Models of the interaction between society and the elderly: a study of opportunities for the social inclusion." The translation from Russian into English was performed by Allen Translation LLC (USA).

Contents

About the Authors

Irina A. Grigoryeva is a professor of sociology at St. Petersburg University and a senior researcher at ITMO University, St. Petersburg. After completing her second doctorate at St. Petersburg University in 2005, she received a special education in gerontology in 2008. She has written 5 monographs and more than 140 scientific publications on social policy study and the comparative theory of social work and, later, on social gerontology and "the old people issue" in modern Russia as well as on the identity of the elderly.

Lyudmila A. Vidiasova is head of the Monitoring and Research Department at the E-Governance Center, ITMO University. After completing her doctorate in sociology at St. Petersburg University in 2013, she has written more than 70 scientific publications on e-participation, ICT adoption and smart cities development, elderly communities on the Internet, and more.

Alexandra V. Dmitrieva is a senior ethnographic field researcher at the ICF "Alliance for Public Health," qualitative research consultant at the World Health Organization, and co-founder of the Support, Research and Development Center. After completing her doctorate in sociology at St. Petersburg University in 2012, she got a job as a fieldwork coordinator in Russia at the Eurasia Program's Open Society Foundations. She has authored more than 20 scientific publications on different aspects of educational ICT and ICT adoption by the elderly.

Olga V. Sergeyeva is an associate professor of sociology at St. Petersburg University. After completing her second doctorate at St. Petersburg University in 2011, she has authored more than 40 scientific publications on ICT adoption, the communication process, media, and everyday life.

Chapter 1
Introduction

This book is dedicated not to old age and old people, but specifically to elderly people between young adulthood and those in the final stages of life due to advanced age.

People do not always recognize this distinction, as the "third" and "fourth" ages have only been differentiated recently. This scientific discovery is attributed to Peter Laslett, who wrote "The third age represents a new historical phenomenon, resulting from successful economic and demographic development and generous social policy. Differentiating the "third age" denotes the appearance of yet another human lifecycle stage... in addition to the classical triad: "childhood—adulthood –old age." Accordingly, old age in this new scheme becomes the "fourth age"" (Laslett 1996). Such differentiation could only arise after the pension system was developed and stabilized in the majority of European countries, America, and other parts of the world. The phenomenon is reminiscent of the way that "adolescence" emerged between childhood and adulthood as elementary school, and now even higher education, developed.

We maintain that two main factors seem to have stimulated the emergence of many elderly-but-not-yet "old" people (i.e. people of the "third" age). First, the presence of a stable pension system in many countries that offers a compensation level of no less than 40% of earnings, and in some cases even 55% and higher. The second factor is rapid expansion of both average life expectancy and the "survival age," i.e. the time following retirement when people live on their pensions and have the opportunity to "reap the fruits of their labor," which used to be possible for only a privileged minority.

In fact, population aging as a whole is primarily connected with decreased birthrate. In other words, comparative aging is possible even without an increased number of elderly and old people. For the time being, it is difficult to tell exactly by how much the number of aged people or the number of people with pension rights has increased, particularly due to the increased number of women entering the labor market. In modern society, any age becomes subject to rationing, control, and regulations, and turns into a designed product, in accordance with which a continuing

© Springer Nature Switzerland AG 2019
I. Grigoryeva et al., *Elderly Population in Modern Russia*,
https://doi.org/10.1007/978-3-319-96619-9_1

life program is built. This especially pertains to "life in retirement," and in many ways eliminates opportunities to integrate older people into "normal" society. At the same time, retirement often entails a decreased social status and welfare level for the elderly population in any country. Of course, this refers to employed workers, not the few "platinum collar" workers or "golden parachute" recipients.

What can be done to preserve accustomed wealth or lifestyle, as pensions may not suffice or may only be adequate when supplemented by savings or investments? What is the solution: to continue working, to "eat through" one's savings, or to resign oneself to a lower quality of life?

Population aging has spawned the global systemic crisis of adapting existing social institutions to society's new age structure. Prolonging the elderly population's ability to work and extending their employment term to preserve independence is becoming one of the most important tasks in terms of older people's interaction with society. Specialists believe that extending educational opportunities and the subsequent employment available for elderly is even more pressing a priority for modernizing society than developing social security (Roik 2012; Elyutina and Chekanova 2003; Grigoryeva et al. 2014). Instead of a "pension policy" or an "elderly social care policy," there is need for a more-encompassing "age policy" or "aging policy."

Meanwhile, stereotypes among both society and experts impose the idea that an aging society's main problem subsists in increasing medical expenses and social services costs for the elderly. These stereotypes arose due the fact that elderly people were initially referred to the Department of Medicine or even Psychiatry when they first emerged as a new demographic group and aging was identified as a social problem (Yatsemirskaya 2006). However, today's biologists and doctors acknowledge that current professional approaches are completely inadequate to explain aging, as there is no cascade of health expenses associated with it and in many cases to be elderly does not mean to be ailing.

In addition, there is obvious disappointment in traditional caretaking strategies, which Western specialists even directly call "disabling." Nevertheless, the government continues to ascribe senior citizens only the roles of recipients and social benefit consumers in need of "protection." Thus, Point 1 of the recently-developed "Action Strategy to Benefit Senior Citizens," declares: "Development of elderly care is a top demand of the time." Women over the age of 55 and men over 60 are, to this day, considered elderly and incapable of work (Proekt... 2015). The above declaration that people are incapable of work and furthermore old (pensions are awarded based on old-age) is a form of ageism (the fear and/or rejection of old age), to which the population has adapted and even learned to receive "secondary benefits."

Gerontophobia and ageism permeate society and make it objectively difficult to view the later part of life as attractive. This is evident from the multitude of ads for makeup to "erase wrinkles" (anti-aging) to mass media's lamenting the elderly as the most unfortunate generation. There is also a directly contrary but just as prevailing trend in society to view today's elderly as "the selfish generation," which we will discuss in further detail below.

The term "ageism" was coined by the English researcher Robert Neil Butler in the early 1960s. He studied how the process of stereotyping and discriminating against people because they are older is analogous to racism and sexism, which are tied to skin color or gender. Contact with elderly people can be considered undesirable as it reminds young people of their own future aging (Butler 1980; Bytheway 1995). In part, this is due to the fact that young people do not have direct experience being old, and subsequently rely primarily on social stereotypes, which are almost exclusively negative. Ageism forms a relationship between generations that does not extend beyond youth's nihilism and the elderly's banal edification.

We believe the problem is that ageism exists in all modern, i.e. quickly-developing, societies. Surely, this is connected with the fact that inter-generational relationships have never been harmonic or idyllic. Notably, "in traditional societies the relationship to elders varied from tender care to the cruelest treatment, even murder. Contrary to the popular opinion about harmony in relations between generations in the traditional Russian community, they were characterized by fairly strong tension, and sometimes turned into outright conflict" (Bocharov 2000: 169–184; Mironov 2009: 138–167). Therefore, it is unhistorical and unproductive to look at elderly social decline in modern Russia as only a moral or ethical problem; it must have a more neutral or rational significance.

There are also several methodological problems associated with studying senior citizens. There is a tendency to romanticize the elderly, attributing to them wisdom or, in contrast, primitivism and senility. Evidently, as sociologist R. Collins attests, perspective depends on the emotional attention center.

What cultural and social factors determine old age? Did they exist before the pension system emerged and marked the beginning of aging in modern society? Do certain cultures exist without the idea of aging, the way that European cultures once circumvented the idea of childhood, if we are to believe Philippe Ariès and other authors?

Meanwhile, senior citizens themselves are chained to the established rhetoric of care, service, and respect without minimal interest in their own opinion about the problems that they allegedly create. "The lack of reflection," writes A. Smolkin, "is especially surprising in terms of gender or nationalism studies' successes... Age remains something external to social theorizing, a subject of biology and medicine, although the above fields' positions are under constant criticism from the social sciences" (Smolkin 2014).

What are we discussing when we talk about aging? Everyone shares the experiences of childhood, adolescence, and adulthood. But what experience is reflected and understood as that of growing old? Aging and old age are abstractions of a multitude of ideas and understandings, which we use to try and understand the "nature" of aging and relationships between the elderly population and other social factions. What experience is understood as aging while a person is alive? It seems that there is no shared experience, but only that which can be lived or observed.

A similar fresh point of view is that of I. Dudenkova: "For centuries childhood was conceived as an object that was rigidly embedded in conjunction with "women and children" or "child in a family." Any attempt to distinguish the scientific

problem of childhood by itself without reference to the professional views of peda-
gogues, the relationship between children and parents, or children and adults was
doomed to failure because of the instant switch from entity to relationships"
(Dudenkova 2014).

What kind of discourse does old age produce?

In modern society, there is a movement away from amorphous and undisciplined
aging present in the rigid, externally-managed pension designation, seniority, and
age framework. It is important to meet the accounting requirements in order to be
part of the "metered man," straight from Foucault.

In public consciousness, childhood and youth are oriented on the future. However,
there is no adequate/authentic orienting point for old age.

- Childhood is a game and a child is Apollo or Dionysus. His or her duty is to be
  happy. Orientation is on the future; "everything is ahead…"

Childhood was reinterpreted in Europe in the 1600s, according to F. Ariès, and
in Russia in the 1880s, according to J. Lotman. It is associated with women, nannies
and governesses, scholarship, reading, romanticizing, and glorifying.

- Adulthood is work and an adult is Prometheus or Sisyphus. His or her duty is to
  be serious, to serve state and country (king and country), and start a family.
  "There is still much to come, to do…" and, in general, "to work is to pray…"
- Old age is relaxation, care, the duty to be poor, lonely, and unhappy… To oppose
  work and health. Is an elderly person required to be either the wise man Socrates
  or the mad man? Is there nothing ahead, just the end? Concentration on life's
  immediacy or on the fact that every moment of life is wonderful?

For the very old, there is no aspiration to overcome the social inferiority complex. On the contrary, usage practices develop, for example, receiving secondary
benefits from acknowledging this inferiority. The exit from control, emerging in the
beginning of 2005 after the "monetization of benefits," resulted in laughable measures: free travel on city transport, and in the summer, to the surrounding areas, as
well.

It is plain to see that life's problems are simplified and schematized, reduced to
low pensions in which politics and mass media play along with one another, romanticizing life's twists and turns: "We are all indebted" vs. "the most unfortunate generation," etc.

The portion of older people who are high-paid specialists and tax-payers are
constantly reminded that from an economic point of view, aging is a luxury.
Governments of many countries today are already concerned about the level of
spending connected with increasing life expectancy. First and foremost, they are
concerned about medical service expenses and long-term care.

Thus, formal generational equality in the normatively regulated "society for all
(ages)" is constantly disrupted even by sociologists in discussing the value of children or risks of aging and, especially, the extreme spending that the elderly population requires – the burden of aging. And while an older person has opportunities to
increase income or diversify one's leisure time in an urban setting, in dying-out

villages there is still "neither hide nor hair of Information Communication Technology (ICT) or television, and kerosene stoves still stand on the windowsills… "as one Russian Science Foundation (RSF) expert noted.

Accomplishing the socio-cultural "senior citizen" project requires new institutionalized practices, which would necessitate all political agents to agree upon an "age policy." And not just in terms of the elderly, as it currently seems that inconsistent and uncoordinated social statuses extend to young, adult, and the elderly. The population should be offered orienting "life models," as one can only cautiously rely on a self-evaluated situation or the evaluation of interested institutions (such as medicine, social work, etc).

However, a policy like "the art of ruling," according to M. Weber, requires consensus in relation to desired results. What result does society need from older people, and what do elderly people need from society: quality of life, social inclusion, or independent lives? How do they relate, and to what extent is a senior citizen the author of his or her own life's story?

For the time being, in modern society there is a noticeable lack of social constructs in which aging and elderly people are studied. The meanings of aging are rapidly changing, although social "sense-making" for now remains. Is it possible to find orientation based on senior citizens' requests, formulating an "aging policy," which could be necessary for the country's future development? There is no simple way to answer this question…

A couple of words about the logic behind this book and the text structure: since we believe that the "risks of aging" are significantly exaggerated, in its Russian-language publication the chapter about social service for the elderly was consigned to the end of the book. In addition, by many modern accounts health depends on maintaining interesting employment and education, and does not linearly depend on age. We followed the book's title, which outlines the sequence "between employment, education, and health." After the Russian-language book's release in spring 2016, we carefully and gratefully followed the opinions expressed about it, both in informal discussions, as well as in published reviews (Parfenova 2016; Bocharov 2017). For us, these opinions were extremely important, particularly those expressed by reviewers at the Springer publishing company about the book's translation by our dear American colleague, Sarah Allen. Subsequently, we not only updated the statistical data, but also changed the chapter sequence.

In the Chap. 1 we establish the intention and main idea behind the book, discussing how little attention sociologists devote to the topic of aging. Visibly, the tradition to believe that children and youth the "hope of society," while elderly are the "burden of society," is still very persistent. We note that in aging conditions, the general Future is the time of the elderly. Accordingly, it is very important to rethink approaches, increase elderly participation in effective, qualified, and possibly more extensive employment, looking back at the calendar less.

In Chap. 2, there are 3 subchapters, establishing the idea of an aging population in the modern world, including a definition of aging; special characteristics of aging in Russia; the relationship of the working and non-working populations; conceptualizing change: from poverty to social exclusion of the elderly. Many statistic materials are introduced, and the demographic processes in Russia are compared with those in several other countries.

Chapter 3 consists of 3 subchapters and introduces the idea of quality of life in terms of age, work, and retirement, including the main approaches to defining the boundaries of "advanced" age; active age, age periodization, and age limits accepted in Russia; increasing elderly employment. The main idea is that modern elderly people leave the labor market later, not because their pensions are insufficient to provide for a decent life, but primarily because they do not want to lose their social status, and want to continue at an interesting job.

Chapter 4 examines the idea of what it means to be old; "elderly" identity as a sociological problem, including the gender approach, and studying elderly identity; age and time, social and individual, generations and conflict; life plans. The chapter considers the impact of "short-term life plans" on aging, and the necessity of constructing a "life scenario."

Chapter 5 provides the idea of specificity of elderly health, the possibility of adaptation medicine, or healing sicknesses, including aging and health loss risks; mass media and older persons' health; "the role of the elderly," "secondary benefits," and adaptation medicine. Death, dying, and loss in old age are discussed, several countries are compared based on the new "quality of life" indicator – the "quality of death" or "death with dignity."

Chapter 6 considers social service for the elderly, including developed countries experience of social services for the elderly compare with Russian social services; interaction of state and community support; characteristics of the elderly service system in Russia. The main defect social services in Russia is in perceiving elderly people as passive subjects to whom the government "offers kindness." Overcoming this relationship is only possible via older people's participation in mutual aid, serving one another, or serving the extremely old, increasing coherence and giving people a way to "see each other."

Chapter 7 provides the idea of education and ICT training practices in Russia for older persons, including research methodology; first interaction with ICT experience, first user practices, and first fears and mistakes connected with mastering virtual space; computer literacy courses, goals and motivation for continued learning; older people's daily interaction with the computer: user practices, difficulties, and reasons for avoiding several actions; elderly interaction and ICT through computer class teachers' eyes.

Chapter 8 provides the idea of Internet space as a platform for studying elderly social inclusion opportunities, including educational resources for older people and their demand; e-government technologies and e-participation tools as new opportunities for older people; discussions on Russian pension system reforms in the mirror of social media.

Chapter 9 describes the emotional experience of old age as a result of media work, including emotional inequality in social stratification theory; studying emotional culture, emotional socialization, and emotional experience; movies as research material; love on screen – 60+: research results and their discussion. Analyzing film characters lets us also talk about the characteristics of "love triangles" in plots, the role of children and grandchildren, the rules of showing initiative in developing relationships.

Chapter 10 provides a conclusion to the whole book. A person retiring today is not considered economically productive, and is thus devalued. We interpret this as modern mass culture's retreat from the principle of anthropocentrism in favor of socio-centricity and labor-centricity. Modern aging society faces the necessity of replacing behavior models in new socio-historical conditions. The strategy of cooperation, acceptance, and solidarity, which recognizes multilinearity, continuity, selectivity, pluralism of individual development due to the subject's unique activity, and the influence of the environment (the general socio-historical and cultural background of all generations), represents the only true step toward a future sustainable society.

We would like to thank our teachers and colleagues for discussions which helped much to improve this book: V. N. Anisimov, A. V. Sidorenko, Z. Saralieva, V. N. Yarskaya, E. R. Yarskaya-Smirnova, M. E. Elyutina, V. N. Kelasev, A. V. Chugunov, A. A. Smolkin, D.M. Rogozin and many more.

References

Bocharov, V.V. 2017. Molodost i starost v traditsionnoy russkoy kulture (o knige Grigotyevoy I.A. I dr. "Pozhilye v sovremennoy Rossii: mezhdu zanyatostyu, obrazovaniyem i zdorovyem"). *Sotsiologicheskie Issledovaniya* 1: 159–163.

———. 2000. *Antropologiya vozrasta*. St. Petersburg: Izd-vo S.-Peterb. un-ta.

Butler, R. 1980. Ageism: A foreword. *Journal of Social Issues*. V 36s2: 8–11

Bytheway, B. 1995. *Ageism*. Buckingham: Open Univ. Press.

Dudenkova, I. 2014. "Detskij vopros" v sociologii: mezhdu normativnostyu i avtonomiej. *Sociologiya vlasti* 3: 47–59.

Elyutina, M.E., and E.H. Chekanova. 2003. Pozhiloj chelovek v obrazovatel'nom prostranstve sovremennogo obshchestva. *Sotsiologicheskie Issledovaniya* 7: 43–49.

Grigoryeva, I.A., L.A. Bershadskaya, and A.V. Dmitrieva. 2014. Na puti k normativnoj modeli otnoshenij obshchestva s pozhilymi lyudmi. *Zhurnal sociologii i social'noj antropologii* 3: 151–167.

Laslett, P. 1996. What is old age? Variation over the time and between cultures. In *Health and mortality among elderly populations*, ed. G. Caselli and A. Lopez. Oxford: Oxford University Press.

Mironov, B.N. 2009. Istoricheskaya sociologiya Rossii: Uchebnoe posobie. SPb.

Parfenova, O.A. 2016. O lyudyah "tretiego vozrasta". Grigoryeva I.A., Vidyasova L.A., Dmitriyeva A.V., Sergeeva O.V. Pozhilye v sovremennoy Rossii: mezhdu zanyatostyu, obrazovaniyem I zdorovyem. SPb: Aleteya. 336 s. ISBN 978–5–906823-45-8. *Zhurnal issledovaniy socialnoy politiki* 14(4): 618–623.

Roik, V.D. 2012. *Mir pozhilyh lyudej i kak nam ego obustroit*. Moscow: Eksmo.

Smolkin, A.A. 2014. Vozrast: V poiskah sociologicheskoj optiki. *Sociologiya vlasti* 3.

Strategia dejstvij v interesah grazhdan pozhilogo vozrasta do 2025 goda. 2016, febr. http://www.rosmintrud.ru/docs/mintrud/protection/203. Accessed 26 Nov 2016.

Yatsemirskaya, R.S. 2006. *Lekcii po social'noj gerontologii*. M.: Akademicheskij proekt.

Chapter 2
An Aging Population in the Modern World

For a long time, population aging was not recognized as a global socio-economic problem. It was first acknowledged as a social problem in the 1980s when it appeared that only populations of the most developed countries were growing old. However, at the turn of the millennium, it became apparent that the populations of post-socialist and developing countries were growing rapidly as well.

Throughout the 20th century, the life expectancy of the general population increased by almost double in most countries. During the 20th century, life longevity – previously belonging to a privileged few – advanced to become a norm for the population majority. Considering that aging is a multi-factor process, one can talk about its specifics in different countries. The process gradually approaches physical old age in addition to age/chronology and is determined by a variety of factors:

- By individual factors, congenital and acquired (gender, age, education, and psycho-somatic factors, etc),
- By micro-social factors (interaction with one's immediate social circle: family, neighbors, friends, and coworkers, etc),
- By macro-social factors (i.e. social, economic, spiritual, cultural, and other societal characteristics).

Since 1950, the portion of the population over 60 years old has increased twofold in developed countries, as we see on Fig. 2.1.

In 2010, this age group extended to one in every five citizens; by 2040 the 60-year line will extend to almost every third person; by 2050 33% of developed country populations and 19% of developing country populations will reach 60 years and over. The tendency for women to dominate the population will continue: women will constitute 55% of the population among senior citizens over 60 years old and 61% of the age group of 80 years and older (Ravenstvo 2003).

In Russia, female population dominance will become even more apparent in older ages (Fig. 2.2).

Over the past 30 years, the average age of the world's population has increased by 5.9 years – from 22.6 years in 1980 to 28.5 years in 2010, according to United

© Springer Nature Switzerland AG 2019
I. Grigoryeva et al., *Elderly Population in Modern Russia*,
https://doi.org/10.1007/978-3-319-96619-9_2

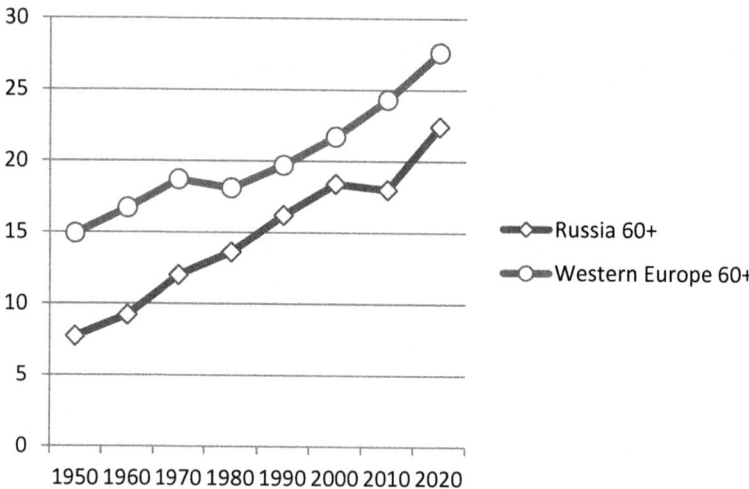

Fig. 2.1 Percentage of total population by age group 60+, both sexes (per 100 total population)
Note: Western Europe: Austria, Belgium, France, Germany, echtenstein, Luxembourg, Monaco, Netherlands, and Switzerland
Source of data: United Nations, Department of Economic and Social Affairs, Population Division (2017). World Population Prospects: The 2017 Revision, custom data acquired via website

Nations research. In Europe, the indicator was 40.3 years in 2010. In developed countries, the aging process has continued for a long time already and occurs mainly "at the top" of the index by increasing life's duration for the older portion of the population; in developing countries, population aging is a newer phenomenon and happens both from "above" as well as from "below" in terms of reduced birthrate.

It becomes all the more apparent that age ranges in terms of growing older are being pushed back to a later time altogether as a significant part of the population faces a longer life expectancy. Medical and hygienic victories, improved working conditions, and raised living standards most places have accelerated society's aging process. We emphasize again: lengthening life expectancy on an individual level inevitably leads to society's aging as a whole, disrupting the count of "checks and balances."

The Russian population is still younger than its central European counterpart by 2.3 years. According to the UN's forecast, the average global age will reach 34.6 by 2040; Europe's average age will be 46.3, while Russia's will be 43.2 years. In comparison with other countries, Russia's population today can be considered relatively young due to relatively low life expectancy. However, aging occurs at a faster pace – if the proportion of elderly (people of 60+ years) accounted for 18% of the population in 2010, then according to the median variant forecast this age group will account for 28% of the population by 2030 (Demograficheskij prognoz, 2013). It is also important to note that the elderly population in Russia includes women of ages 55 and up, which causes some confusion in calculations. However, if the retirement

Female   Male

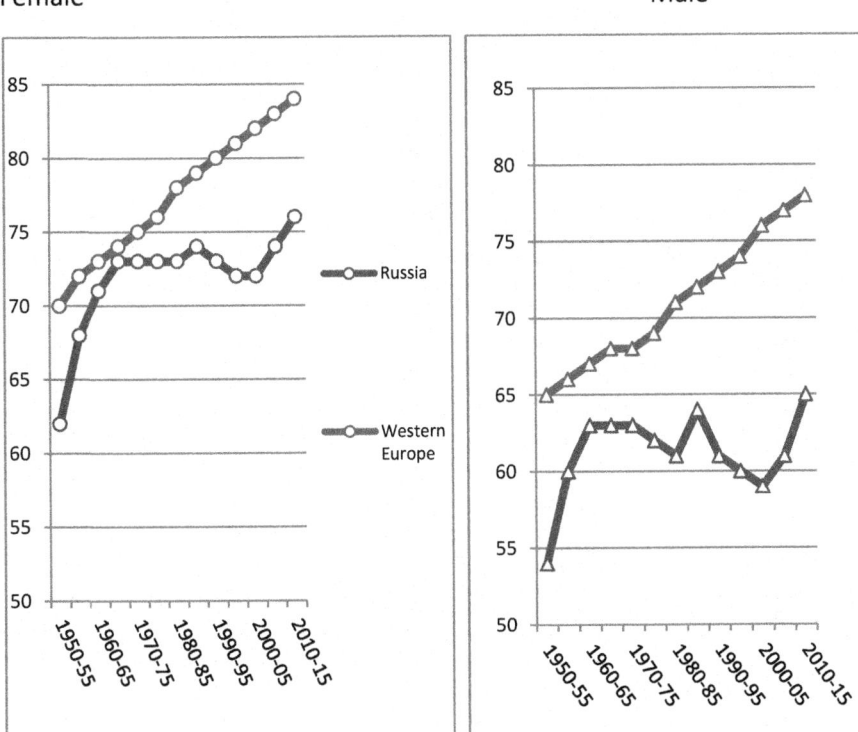

Fig. 2.2 Life Expectancy at Birth
Source: United Nations, Department of Economic and Social Affairs, Population Division (2017). World Population Prospects: The 2017 Revision, custom data acquired via website. http://esa. un.org/unpd/wpp/DataQuery/

age were to even out between genders or to increase, the elderly population proportion might grow more moderately.

2.1   Defining Aging

Although aging has already become a subject of special sociological and philosophical study, it remains more-or-less acceptable for the socio-humanitarian sciences to define the process of aging and the elderly.

In the most generalized sense, it is the broad designation of a group phenomenon that leads to reduced life expectancy with age on an individual level and to an increased number of elderly people in the population on a social level. This definition clearly issues from a demographic, and not a social standpoint.

The next definition is biological, based on the theory of organisms' adaptation: aging is a multi-sectional process that unavoidably and naturally increases over time and leads to an organism's declined adaptive capacity and increased probability of death. Through the processes of growing older, age-related changes to the organs and tissues accumulate and combine with those caused by external influences. The conventional changes caused by aging are physiological processes, which although not primarily diseases, in the majority of circumstances still limit the organism's functional capacity and decrease its resistance to various harmful influences. However, this is just one of the, albeit typical, perspectives on the aging process. There are a wide variety of other views, which are traced in the works of famous Saint Petersburg gerontologist, V. N. Anisimov (Anisimov 2008).

Historical anthropology has developed its own definition of life periods, combining chronological age and family status. Accordingly, a woman who has grandchildren is a grandmother, independent of her own age, and has transitioned into the older group. A man who in not married on time according societal standards is considered "young" even if he is 30 years old. Society is structured specifically in reference to age "rules," dictating when life events should occur.

From anthropology and other social sciences comes the concept of "age-homogenous groups," and in modern society this construct works well for short-term life periods, like "teen years" and "youth." However, in modern society's longer life periods, people of the same age are becoming more heterogeneous in terms of socio-economic status, so age explains little. Aside from this, "life-long learning" turns people into eternal scholars or students, and increasingly frequent third and so forth marriages allow us to call people of any age "newlyweds" or "young couples."

Economists talk not about old or young people, but about people of work-capable ages and incapacity for work. As a result, older people are considered to be "past working age." Based on more expansive traditions than retirement legislation, economists consider work-capable people to be those ages 16–72. However, even in such a wide range, economists notice that the dependent load (i.e. the burden of those unable to work on those capable of work) is currently growing, but due more to the elderly population section and not to children. This is visible in the "old" states of Europe (Eurostat 2013). At the same time, there are still countries where the burden grows largely due to children as well as the elderly, as, for example, in the cases of India or South Africa.

The definition of aging doesn't meet the requirements of any one science: "old age is a bio-physical and socio-historical concept with conditional and changing limits on the various stages of historical-evolutionary human development and in different eco-populational and social groups." Even by adequately highlighting differences, we still do not capture the "essence," or in this case, we remain within the limits of demographics.

In developed countries, aging is somewhat more pronounced in the female portion of the population, which is explained to a certain extent by the higher unnatural mortality rate of men and their premature aging. However, in other countries where there is a high maternal mortality rate and women bare many children the situation is more complex.

In contrast to discussions about aging risks, in recent years, discussions are sparking up about "active aging" or "active longevity" (along the lines of Butler's theory on active and passive longevity), particularly: "active aging denotes the aging process of a person possessing good health, feeling himself to be a competent member of society, receiving more full satisfaction in productive activities, large independence in daily life, and acknowledging himself as a citizen actively involved in social life. (Eurostat 2013).

We emphasize that from the "European" point of view, aging should be active and independent. A. Sidorenko, an expert at UNECE, explains what this actually means: "Independence: preventative healthcare; physically-accessible transport provision; improved habitation environment quality.

Occupation: creating conditions and better opportunities for older workers.

Social Life Participation: overcoming social isolation; broadening active participation in society; supporting volunteer movements; supporting individuals supplying informal care for the needy" (Sidorenko and Zaidi 2013).

The aging population trend is irreversible, although an "equilibrium point" exists. As the history of France demonstrates, for example, depopulation began at the end of the 19th century due to decreased birthrate, although population size has stabilized over the past 40–50 years. Migration processes, becoming more intense today, are making a significant contribution to the demographic balance. However, younger people's predominance in the population's makeup, characteristic of previous historic eras, is unlikely to occur again. It is worth noting that youthful predominance is an extremely wasteful type of society development model as it requires enormous investments in education and new job creation. This outcome also looks negative in the context of recent decades' ecological alarmism….

Therefore, population aging signifies the opportunity for more intensive socioeconomic development. Growing older as a new phenomenon in historical-evolutionary human development accompanies such evolutionary transformations as sped-up physiological maturation (acceleration), longer childhood (primary socialization), and education (secondary socialization). In parallel, many researchers note a slow-down in terms of physical, spiritual, and social aging.

2.2 Special Characteristics of Aging in Russia

Although old age sets in earlier in Russia (we will explain why later) and the country's mortality rate is much higher than in Europe, North America, and Japan, there has been a significant decrease in the mortality rate over the twentieth century. Thus, the expected average Russian life expectancy in the beginning of the twenty first century was 65 years. This figure is practically twice that of 100 years prior. However, the average life expectancy depends, first of all, on infant and child mortality rates, as these age groups in particular "even out" the index. The childhood mortality rate in Russia truly is decreasing, which has decidedly boosted the average life expectancy within the first decade of the 21st century to almost 72 years (Average Life Expectancy, 2014).

The specifics of Russia's demographic situation consist of the birthrate level, which has gradually fallen to match developed countries' indexes, and the mortality rate, which has remained high like in developing countries. The Russian population's birthrate (per 1000 thousand people in the population) from 1920 through 2005 has decreased as follows: 1940–6.1%; 1960–5.6%, 1970–3.6%; 1980–1.3%; 1990–2.5%; 1995–4.1%; 2005 + 1.5%; 2005 + 0,7%;. Statistic data demonstrates a long-term trend towards lowering the birthrate, which is seemingly uncorrelated with the population's quality of life. Essentially, we witness a decrease in the birthrate that is independent of the numerous and very abrupt social, economic, or political changes that are characteristic of the country's development.

The combination of low birth rate and high mortality rate has induced a rapid decrease in overall population size, which is partially counterbalanced by migration growth, although both the Russian population and specialists view this movement ambiguously.

Consistent with international criteria, a country's population is considered old if the portion of people over the age of 65 exceeds 7% of the population's structure (criteria of J. Beaujeu Garnier and E. Rosset's Demographic Aging Scale). Russia's population might be considered as such since the end of the 1960s (specifically in relation to the Russian Federation, as the USSR as a whole had a younger population structure thanks to its Central Asian republics).

The paradoxicality of Russia's situation relates to the fact the country has experienced an increase in the survival age of the elderly portion of the population against the backdrop of the slow growth of the population's average life expectancy (in comparison with developed countries) as a whole.

The high rate of unnatural male mortality (in particular, ages 45–55) is countered by rapid growth in the number of women (mainly over the age of 80). In older age groups, the gender disproportion continuously grows and by the age of 70 (and up) approaches 2610 women per 1000 men (Safarova 2008: pages 15–16).

The mortality rate in Russia is among the highest in European countries (15 in 1000 people annually; on average in the EU the rate is 6.7 in 1000; 11 in 100 in Romania and Latvia; and fewer than 6 in 1000 in Sweden) (The European 2007).

"Washout" of the working portion of the population (particularly male) under low birthrate conditions (insufficient for simple population reproduction) undeniably affects not only the population's average age, but also the far more important economic position of families and the opportunities for a country's socio-economic development. Early mortality slows the process of a population's comparative aging, but essentially has no positive effects. It is vital that conditions be improved "from the top" of the index, i.e. increasing life expectancy by extending the survival age, particularly among men.

Nevertheless, the renowned demographer Anatoly G. Vishnevsky believes: "in the first decade of the modern century the demographic burden in Russia was at its lowest in the country's entire history (in terms of both elderly people and minors).

In particular, firm requirements for increasing retirement age emerged in this time as the only way for the state to make ends meet. Unfortunately, in Russia the opportunities created by that very development process are yet used insignificantly, and now have been almost entirely exhausted. For example, the expected remaining lifespan of a 60-year-old man in France is 22.5 years, while in Russia it is 14.4" (Andreev and Vishnevskij 2014).

This is very bad from the viewpoint of life's value, but indicates that at the same age as people retire in Russia (60 years), the French government must support a retiree for 8.1 years (1.6 times longer than in Russia). The decision to raise the retirement age has already been adopted in France. However, the importance of adopting a similar policy in Russia is still widely disputed.

As a result, demographic data from recent years allows us to draw conclusions about trends in relation to increased number of old people and population aging. This pertains to growth of the extremely old segment (age group 85 years and up) in the older population, which is determined to be an at-risk group; the growth of the number of women amongst aging and older people; the increase in women's life expectancy by approximately 12–13 years in comparison to men. This analysis leads to the average family's life cycle ending in a long period of widowhood and female loneliness.

2.3   The Relationship of the Between Retirees and Employees

Against the backdrop of social focus on the environmental, energy, and resource problems facing mankind, the world community perceived aging as a new threat to society's sustainable development. However, we venture the guess that this occurs because the modern person is accustomed to thinking in terms of risk. Expectations previously existed that improving the quality of medical help, living standards, nutrition, and growth of general welfare and benefits in European countries would involve demographic growth. No one suspected that global economic development would have negative effects: in poor and impoverished countries it is impossible to curb population growth through regulatory means due to traditions of high fertility (in India, Pakistan, Indonesia, and Egypt, for example), while the birthrate in wealthy countries sharply decreases while the average lifespan sharply lengthens. The "Central-European risk" will not be the population's extinction, as was suggested at the beginning of the twentieth century, but rapid aging and "pressure" from the non-working segment of the population on the working half (Fig. 2.3).

The age structure of the modern American, European, and Russian populations is also deformed as a result of the post-war "Baby Boom," leading to the multitudinous generation born in the second half of the 1940s and 1950s. Recently, this generation reached retirement age, creating the so-called "retirement bubble." Moreover, the generation now beginning its working life is significantly smaller. Therefore, it is more apt to talk about the "aging crisis," rather than about the crisis of existing pension systems.

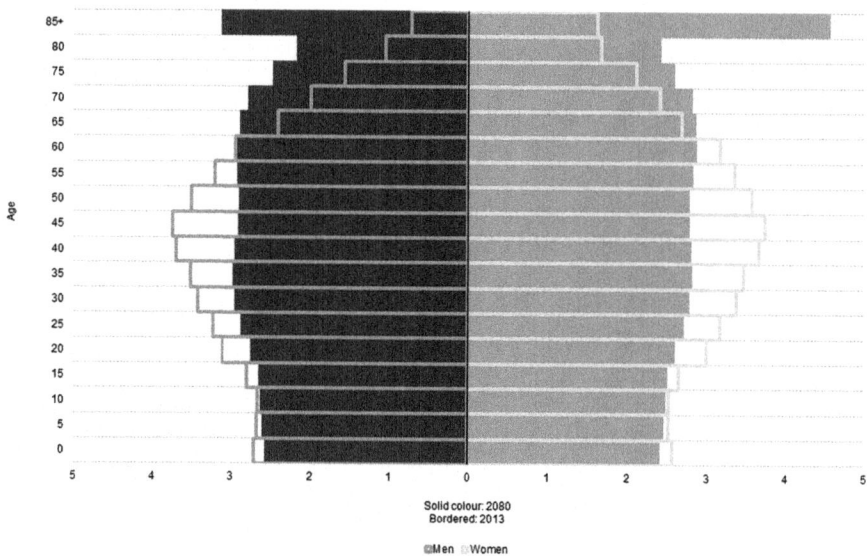

Fig. 2.3   2013 provisional. 2080 projections (EUROPOP2013)
Source: Eurostat (online data codes: demo_pjangroup and proj_13npms)

However, there is reason to believe that the situation in Europe will soon even out. At least, long-term demographic tendencies seem to indicate such a possibility, looking at Fig. 2.3.

At a glance, the diagram shows the modern (2013) structure of the European population with the demographic waves up through the year 2080, which have a high probability of being replaced by a "columnar" structure with minimal fluctuations and good opportunities for calculating economic burden.

One may recall that the demographic structure of pre-revolutionary Russia was also foreseeable, although more traditional in form, which the diagram clearly shows (Fig. 2.4). Visibly, the childhood mortality rate was high, and this could be considered the most critical demographic problem. But in "adult ages," the population gradually and predictably declines.

In terms of today's situation, demographic losses from the World War II summoned the following waves, which were exacerbated by government attempts to intervene in the demographic process. Moreover, stimulating the birthrate in the 1980s partially conditioned the sharp decrease in the 1990s due to the displacement due to the calender births of second children. But this miscount repeated, and the Birthrate Stimulation Program "Maternity Capital," which began in 2007, coincided with the rise in birthrate due to the rise of a demographic wave (Fig. 2.5).

However, unfortunately in Russia the situation is less predictable due to state attempts to "stimulate the birthrate" without adequate measures to improve the population's health and decrease mortality rates of work-aged people. This is well-exhibited in Fig. 2.5.

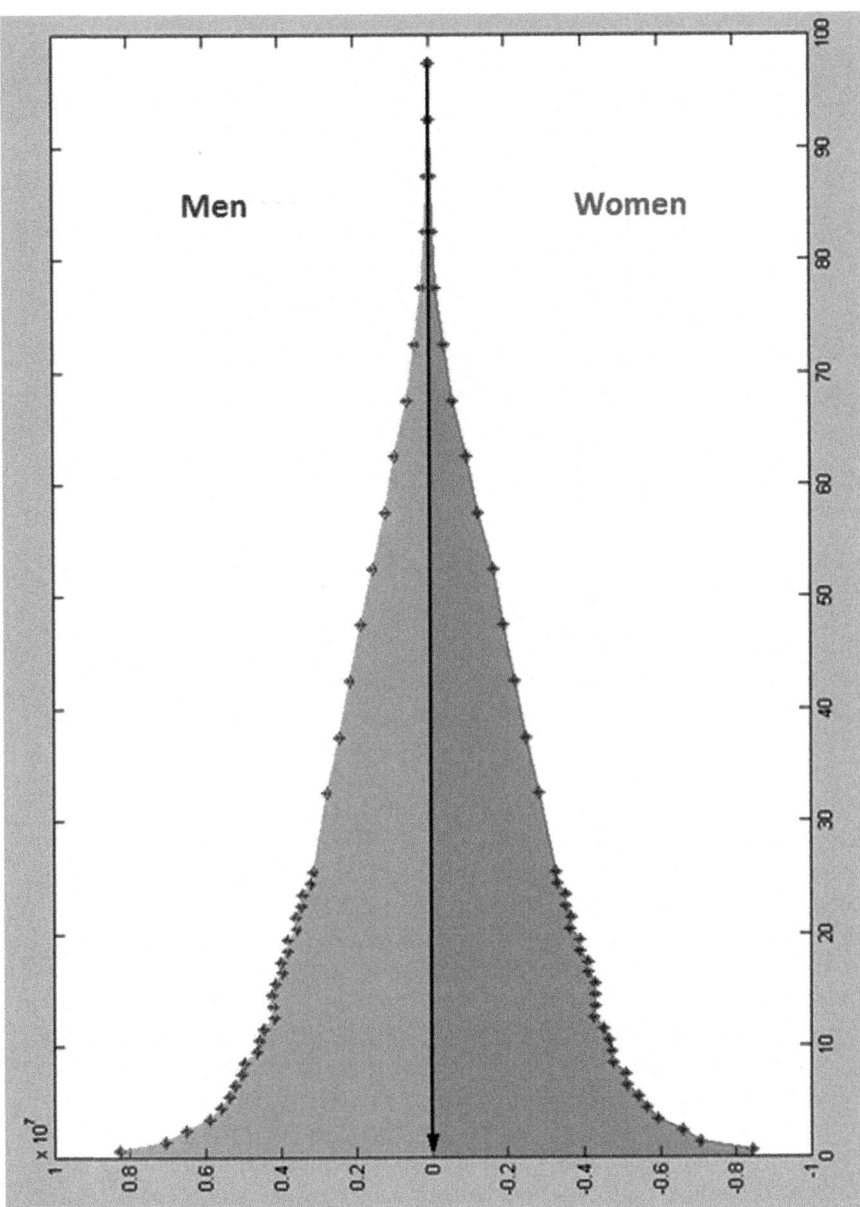

Fig. 2.4 Demografic Structure of Russia Population on 1891
Source: http://demografia.narod.ru/

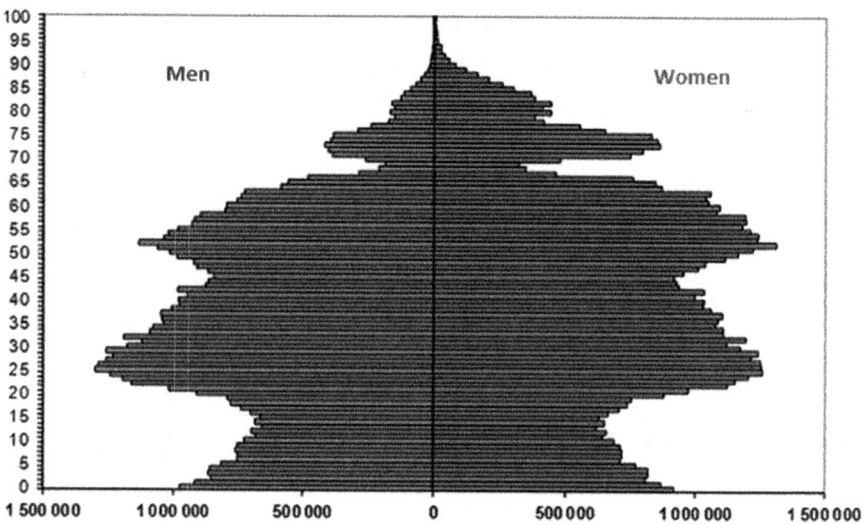

Fig. 2.5 Age and sex pyramid of the population of Russia at the beginning of 2013. Source: Demographic every year Review. 2013
http://www.demoscope.ru/weekly/2014/0585/tema02.php

Thus, demography indicates a repeating situation of the "demographic crossing," which occurred in the 1990s (Fig. 2.6). In recent years, the population's income is decreasing, social stability is replaced by strain, and the number of mothers of a reproductive age is "obscenely low." In order to increase birthrate, people need to see good outcomes again, but to do so depends on general government policy, experts believe. In connection with this, the Government of the RF extended the "Maternity Capital" Program, while President V. V. Putin signed a law on monthly disbursements for first children (Federal'nyj zakon 2017).

In the future "demographic waves" will not cease to rock the possibility of a socio-economic prognosis, requiring a flexible policy, alternating increased places in kinder gardens and schools with more social services accessibility for the elderly. Waves on Russia's age and gender pyramid have negative repercussions for the country's socio-economic development. The obvious repercussions are wavering demands for educational facilities and health care services. The less obvious repercussions are wavering applications in the labor market and the level of demographic strain," (Maleva and Tindic 2014). A study by the European Commission in 2005 demonstrated that there are also demographic repercussions: the generations that lived through the Baby Boom at a young age demonstrate in later years a lower birthrate, resulting from problems such as employment, housing, and private education that they experienced growing up. (European Commission 2005).

Pensions in such a situation should either be constantly recalculated, or should be more connected to age, and not to former labor contribution. Moreover, this situation discourages young people from earning a pension, since social deductions increase in periods of large elderly burdens. Meanwhile, pension age increases. We

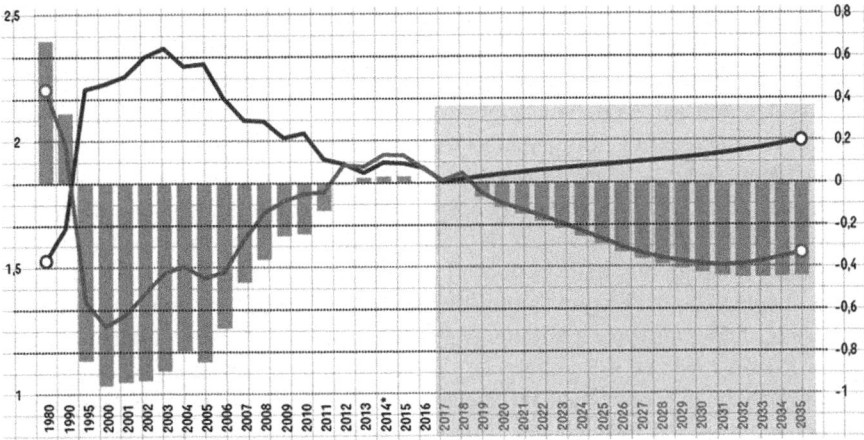

Fig. 2.6 The demographic cross of Russia. (Births are indicated by the blue line, deaths by the red line, and the natural rise/decline by the pink bar), millions of people

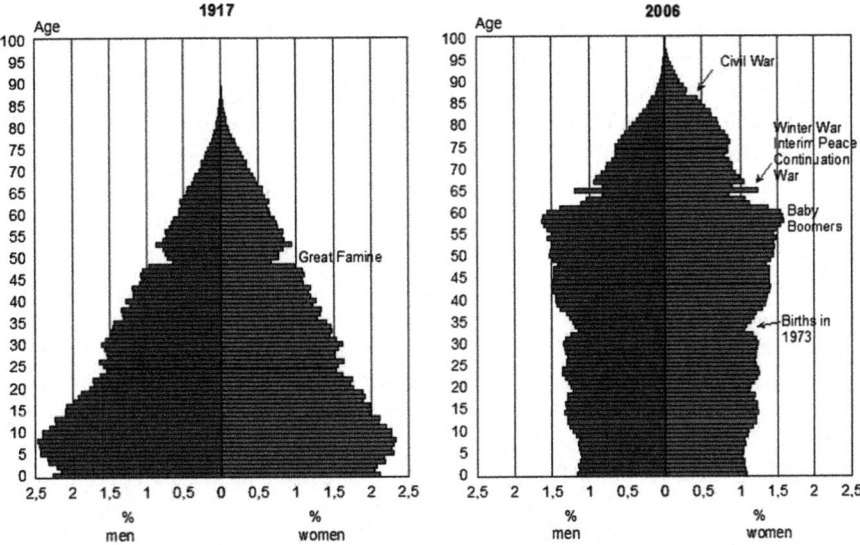

Fig. 2.7 Changes in the demographic structure of Finland's population over 90 years from 1917–2006 (from the traditional pyramid associated with high birthrate and high mortality rate, to decreased birth-rate and mortality rate and increased lifespan)
Source: http://countrymeters.info/ru/Finland

emphasize that this situation requires a flexible resolution, while politicians have no such experience.

There is yet another example: comparing changes in Finland's population structure from the time of gaining independence until the near present shows that changes may be more gradual and oriented on population conservation. We can see at Fig. 2.7.

Changes in the demographic structure of Finland's population. Obviously, judging from Diagram 8, the modern population structure in Finland resembles that of Europe in general, where population loss at different ages is minimal.

In this day and age, economic globalization and the precarious nature of the workforce (i.e. increased flexibility and a rising number of part-time positions) contribute to aging, in terms of the growing number of "economically dependent" people (Standing 2011). Consequently, an ever-greater tax burden falls upon full-time employees. For European women, this new situation centers around part-time employment, outsourcing, and "poor work environments" (Shershneva 2014). As such, the percentage of the population (the majority of whom are women) forced into these unstable part-time positions is 39.4% in England, 9.7% in Finland, and 2.9% in Russia (Women's Employment 2014). With modernization, society has allegedly abandoned former legal and social employment prerequisites: mass unemployment is integrated through new forms of diverse part-time positions in the employment system, accompanied by a variety of risks and chances, from the quality of workers' lives, to opportunities for pension coverage.

The world financial crisis has added "fuel to the fire," as many Pension funds have lost investments in recent years.

This context has given way to the idea that pensioners are becoming an encumbrance on society…

Hence, the debate about the "egotistical generation," too numerous, whose members lived in the golden age of the welfare state in the West or the rise of socialism in the East, and are consequently spoiled and urging too steep demands on society and government (Gilleard and Higgs 2014). This generation is considered responsible for limiting the "life prospects" of the younger portion of "adult society" in this time of global economic decline and states that are increasingly more socially oriented on social costs.

It is valuable to consider these issues in Russia, where there are significantly more pensioners than the elderly, due to the low pension age. Naturally, it is also imperative to change the accepted view of aging as an increasingly inferior life stage, as losing the normal ability to function grows linearly.

This view is unproductive and one-sided, reduced to a medico-biological and demographical focus, overwhelmed by arguments about the inevitable build-up of pathologies and diseases. Then again, this is the natural consequence of counterproductive relations between society and one of its institutions, i.e. medicine which in turn cannot reconsider/find a role in preserving health, not just curing diseases.

2.4   Conceptualizing Change: From Poverty to Social Exclusion of the Elderly

In the post-modern era (i.e. that following the "New Age" as the era beginning approximately in the 1970–80s), human practical activity and social interactions are starting to be considered in sociology as derived from processes of communication, as well as its content and organizational forms. In the "space of culture and communication," sociologists now examine processes, which in the 19th century were considered to be determined primarily by economics, specifically the processes of class, group (stratus), and caste differentiation in society. Thus, as Karl Marx believed, the relationship to means of production is a basic class and stratus-forming principle.

Today, however, a significant number of social thinkers acknowledge that productive relations in general are derivative and are determined by including active social figures, people, or groups, in reproducing existing knowledge, communication, and their cultural framework. Social and cultural frameworks in society are being transformed. Thus, concerns arise with regard to new types of social exclusion – the exclusion of those whose knowledge has become outdated or who lack sufficient skills to be included in more complex communication spaces than television.

"Inclusion/exclusion" has become a basic criterion for structuring modern society. Under the umbrella of "social inclusion," are people who understand the process of integration and entering society. Thus, another axis is added to the usual hierarchical class-group societal structure: inclusion/exclusion. According to P. Abrahamson, "the former class stratification, segmenting people into vertical layers, is gradually combined with the horizontal differentiation into "insiders" and "outsiders" (Abrahamson 2001: page 158). A. Touraine even more drastically emphasizes the transition from a "vertical" society to "horizontal" one in which it is "much more important to understand not whether people are at the bottom or top, but if they are in the center or periphery" (Touraine 1991).

As a result, problems of integrating senior citizens into the socio-economic sphere, as well as increasing level of elderly social activism and functional literacy, become all the more important today. We stress the importance of this task because the focus on the low pension problem paints an inadequate picture of the elderly community's economic position. This position should be evaluated with consideration of their accumulated assets, both immovable and moveable, as well as their social exclusion level. Moreover, this focus point is ageist in essence, as it reduces senior citizens' needs exclusively to the subsistence level and ensuring its implementation.

The term "social exclusion" was introduced by R. Lenuar to represent situations pertaining to socially unprotected population groups, such as orphaned children, the disabled, the mentally or physically handicapped, unemployed, homeless, and other "wasted people" (a term coined by Zygmunt Bauman). This approach to exclusion focuses on wealth distribution in society, access to its resources, and, in consequence, insufficient satisfaction of people's needs.

This interpretation of social exclusion resembles the concept of poverty but, as Abrahamson writes, poor people were never previously excluded, as they constituted a large part of the population. In contract to the socially excluded, the poor were integrated into society and, despite low living standards, did not disappear from "social networks." Inequality and poverty became interpreted as gateways to social exclusion in modern societies, which are built on the norms and values of equality and equal human rights. In traditional caste and class societies, the principle of hierarchy provided every individual with a designated place in the public whole, meaning inclusion and identity, the aggregate of individuals constituted a symbolic unity.

Hence, in European medieval communes (local communities), orphaned children and lonely elderly people remained a part of society, which found opportunities to feed and accommodate them in "foster" families.

Today in Russia's rural areas, especially in isolated regions like the Archangelsk area, a model is developing to form "adoptive families for the elderly" if an older person is alone and the family has no grandmother or grandfather.

Poverty and begging were socially condemned in the West, in contrast to Russia, which has been called "poor-loving". However, this love for the poor was also a form of social acceptance, if not inclusion, working towards a symbolic unity of impoverished and those who served them. In any case, the question about elderly being "other" was not mentioned. This presents the picture of social interactive connections back to to Peter I, who posed the question of eradicating poverty and punishment of both the mendicant and those begging for alms.

Relying on modern Russian experience, N.V. Tikhonova comes to the conclusion that a "certain income level (as well as the presence of other resources, such as the "right" connections, ultimately also affecting total income), allows a person to find an alternative to conventional integration mechanisms in society and to avoid social exclusion even when of discrimination is present" (Tikhonova 2002). Thus, the main difference between the concept of exclusion and the concept of poverty is shifting the focus from inequality in welfare distribution and arguments about the fairest distribution to equal civil and social rights. These rights might be limited by lack of access to social integration institutions and upwards mobility, primarily to education.

But her point in the Russian context there is something special. Education and professional qualification, as the main social mobility channel and for transforming poverty creation, have stopped working over the past few decades, as many authors note (Konstantinovsky 2008). Although it is still true that they play a large first-step role in preserving senior citizens' social-professional status and opportunities to continue professional activity and, subsequently, continue to be included in their usual social spheres.

"Exclusion" in socialization processes. The concept of social exclusion is connected with socialization processes. Social inclusion/exclusion practices take place

on an individual level in the process of socialization, as well as on a communal level in terms of the government as a whole. On the individual level, "exclusion" refers to proper socialization disruption, extrusion, and exit from the margins of normal community leading to long-term consequences.

However, socialization in an endless process: "it is not easy or is even difficult to be young, adult, and elderly, but this is not so much determined by age as much as by specific features of the social environment in which socialization processes affect every age" (Grigoryeva 2006).

At the same time, the main approach in relation to the elderly is one in which aging processes are treated as irreversible desocialization, for example, in the theory of liberation/division championed by American researchers Elaine Cumming and William Henry. They proposed that "in the aging process elderly people become alienated from those who are younger; aside from that, a process takes places liberating elderly people from their social roles, meaning the roles associated with work activity. This process of alienation and liberation is derived from the social context in which the elderly people are present. This might be considered one of the ways of adapting to limited opportunities and accepting the inevitability of impending death. In some respects, elderly people feel "more free," as the necessity to work and participate in social life no longer looms overhead. According to the theory of liberation, in the social aspect the process of alienating elderly people is unavoidable, as the duties they fulfill at some point must be passed on to younger people who are capable of working more productively" (Smelser 1988: 374).

However, this theory obviously exaggerates the positivity of the process of employment liberation, as it promotes elderly isolation and discourages adaptation. Yet, in modern society, the roles associated with labor are not necessarily tied to a specified age period. All the more often, highly-qualified specialists work as long as they want or are able to, as proven by a growing trend in different countries. Thus, an exhaustion/liberation from labor roles crisis is impossible.

Nevertheless, "new adaptation strategies are needed in the final stage of life that are conducive to mastering the socio-psychological mechanisms that block negative effects in an elderly person's critical life moments" (Yelutina 2002: p. 130). Transition to life in retirements is potentially a turning point in the terms of individuality and contains the valuable opportunity for personal growth through full realization of potential that was previously undiscovered.

The principles of interaction between older people and society, as recommended by international organizations, are also fundamentally changing in recent years. Mid-twentieth century international documents accentuated the meaning of such social rights as the right to a pension, but today the primary focus is on the importance of manifesting senior citizen's rights in the field of labor, maintaining occupation, and participating in feasible employment.

In the "United Nations Principles for Older Persons," (Principy 1991), the first principle relates to retirees' independence. Elderly people should have the opportunity to work or pursue different types of profitable occupations; they should have the opportunity to participate in specified periods and forms of ending labor activity; they should have the opportunity to participate in relevant educational and professional preparation programs. In order to continue employment, it is imperative to create a mechanism for voluntary retirement and to provide flexibility in setting a retirement age. The "Madrid International Plan of Action on Ageing" also attests to the importance of guaranteeing senior citizens' right to employment (Madrid International 2002).

In general, elderly people's level of adaptation to the aging process and retirement depends on a multitude of factors, beginning with motivation and ending with the nature of the employment they held in earlier life stages. Preserving activeness and the diversity of possibilities relating to life is seen as a positive social inclusion factor. From this life position, people are open to changes, which they perceive not as a threat, but as a test of one's potential.

According to Bernice Neugarten, "if by the time approaching old age, a person has held a multitude of different roles, he will find it easier to endure the loss of the roles he performed in the past" (Neugarten and Hagestad 1979: p 38). At the same time as passive life position and role narrowing turn into the desire to shift the burden of responsibility on someone else, institutions and educational entities seek to resolve all related problems including, ultimately, social exclusion and desocialization.

The resource-based approach. Similar to exclusion processes, inclusion processes occur on various levels. Stepping back from the connections between inclusion processes and the concept of continual socialization, which we discussed earlier, inclusion processes seem to be distributed in at least three social dimensions, as N. E. Tikhonova comprehensively describes. She thus determines three levels of inclusion or three measures of social capital: (1) inclusion in an informal setting with friends and acquaintances, (2) individual participation in social organizations/associations/informal communities, (3) the presence of connections with individuals from high social strata, to whom, if necessary, one can turn for some kind of help.

The researcher emphasizes the difference between an individual's real social capital and resources similar in function. This difference consists of the fact that, in contrast to resources, social capital can be incorporated in economic capital, thereby affecting an individual's economic position. Or at least the presence of connections that can be activated when needed eliminates unnecessary financial expenses, i.e. preserving the economic resources an individual has already amassed. These connections can essentially simplify access to the inaccessible, for example to power resources, or help maintain "social face," i.e. avoid uncertain situations or stigmatization (Tikhonova 2002).

Obviously, inclusion processes are in many ways connected with social capital distribution among different groups and the opportunities for some individuals to "borrow" this capital from those in more affluent groups. Although in life we do often still observe interactions and resource exchanges between individuals from similar social groups, which support their production/reproduction.

In particular, these social transfers take place between different generations in a family in both Russia and in the West, from parents to children or from children to parents. Because of this, the tie between socio-economic capital appears all the more profitable, especially in countries where economic and bureaucratic relationships can be built on trustworthy kin/clan bases.

"Inclusion/Exclusion" in Consumer Society. Although the conditions and prerequisites of inclusion and exclusion are closely connected to one another, socialization is considered in Russian scientific literature to be far better understood than desocialization. This speaks to the fact that these concept pairs still represent difference processes. Socialization is studied, while inclusion – i.e. the instrumental side of socialization – is not. The technical side of the question is far more widely studied, as it concerns excluded people whose physical and other limitations interfere with leading a normal life style. Thus, the results and prospects of transforming urban space is currently analyzed to consider the needs of handicapped, blind, and elderly people, and so forth. The concept of a "friendly-oriented city" has emerged, principally in relation to the large elderly community.

In modern society, the established economic inequality measures inevitably predominate are supplemented by less-well understood and unambiguous measures. Where previously there were "rich" and "poor" measures, now there are "poor lifestyles" and "rich lifestyles," between which emerge even more diverse lifestyles, uses, and other separate groups as well as even individuals.

Socially-approved and condemned lifestyles are appearing and becoming a factor of inclusion/exclusion. The multitude and multidimensionality of modern society, which is developing simultaneously inside and outside of the established social order, create specific difficulties in the need or desire to somehow structure or classify it. Any category is ultimately no more than a social construct, the fluidity and flexibility of which depend on definitions that are accepted and considered comfortable at a given moment. As Bauman notes, that which seemed satisfactory yesterday becomes dangerous today; that which was considered normal requires (medical, social, and state) intervention today (Bauman 2004; p. 87).

Stylistic Features of "Inclusion/Exclusion. " Exclusion processes more frequently are characterized by change from one life style to another; individuals excluded from one relationship or activity are included in others that do not necessarily stigmatize or negatively influence one's preexisting social position. Instead of a stagnant "poverty culture," the abrupt change from a production society to a consumer society gives way to "lifestyle of the poor".

This occurs, according to several authors, not only out of need and truly low incomes, but often out of convenience or even enjoying the feeling of being "disadvantaged" or "uncoddled" by the caring hand of the state. In order to satisfy the needs and demands of such groups newer social services continuously emerge, striving essentially not to help, but to strengthen existing self-definitions, self-awareness, and the stylistic characteristics needed in this population group. Much has already been written about the "secondary benefits" of such self-stigmatization, and we will also return to this topic several more times.

New methods for exclusion are also flowing in on the wave of ongoing changes. Lifestyle, which is organized on details that frequently contradict each other from a normative point of view, can contain numerous reasons for exclusion. Hence, professional accomplishments and high educational level without the corresponding high consumption style, for example in terms of food or clothing, might betray an individual for the person he is, not the person that he wishes to seem (Bourdieu 2005). This hardly excludes him from the professional community, but might become a way to seem "better" and can contain prerequisites for exclusion on an inter-personal level.

Unluckily, inclusion/exclusion is not just a question of the interpersonal relationships in which each is bound to differentiate various phenomena or other individuals based on personal psychological projection. As previously mentioned, the government plays an essential role in these processes. The logic of state actions, with rare exception, has not kept pace with the logic of social development, almost always the state's influence is universal but not differentiated.

The concept of the so-called "Welfare State" and actually existence of developed welfare states beginning in the early 1990s are in a constant state of crisis. This crisis, as many authors assert, is connected with the social state eroding due to increasingly dominant neoliberalism ideology and the state's exit from the field of important socio-political decisions.

The victims of this process are not just the poor, but also the "middle class of wealthy workers," and now frequently the elderly who are becoming "too numerous" for state care (Taylor 2010: p. 36). In various countries, one can observe increasingly active reallocation of incomes and tools to benefit big business (trickle-up effect), the contrasting "trickle-down effect" promised by liberals, or rewarding senior citizens for "labor participation," to which all social democracies from Sweden to the cling.

Consequently, the only way to preserve or improve the social exclusion situation is through inclusion practices, which society can and should practice though not necessarily through state support. At this moment, the Russian state's actions to resolve social problems are more directed towards finding new reasons and nominations for exclusion, as well as on the latest isolation of deviant individuals or undesirable organizations from normative society. The paradox lies within the fact that the numerical relations and quality of connections inside excluded societies that formed on the basis of inclusion in nonstandard practices constantly grow and strengthen, while the durability of the existing social structure becomes less certain.

Particularities of Inclusion/Exclusion in Russia. It is well-known that there is a large gap between poor and rich in the Russian social structure, and furthermore this situation is gradually changing towards differentiation by "horizontal" rather than "vertical" social characteristics. Accordingly, the established inequality categories are gradually supplemented by new categories as the structure of modern society changes. In this context, age joins the ranks of factors explaining social inequalities.

Even in the complex years of the 1990s, several specialists noticed that the elderly people's situation, in the face of obviously decreased income (pension), often does not fit within the margins of the category "poverty." Subsequently, the level of elderly housing provision, including one's own apartment or house, was significantly higher than amongst young people (Housing economy... 2002). In addition to the apartment, many elderly people possessed dachas (country houses) in various degrees of development and some considerable personal property.

Nonetheless, no social workers' efforts could consistently persuade older people to capitalize on their property, for example, in the form of renting out one of their apartment's rooms or exchanging one's current apartment for one of a smaller size. A significant portion of property was old and objectively difficult to use for capitalization, although still in daily use. The situation itself concerning numerous "black deals" with real estate in which elderly people were either evicted to unfit housing or even killed, attests to the risks of solitary living in old age as well as possessing property in excess.

The retirement system's development made life in old age more comfortable but broke the tie between elderly and adult generations; between retirees and workers. As a result, social exclusion has become an aging risk in modern societies within the accepted frameworks of the pension system and the prevalence of urban lifestyle, wearing away at the mutual support system.

From the Russian government's example, we observe the phenomenon of the "heartlands" which, according to N. Zubarevich and V. Ilyin, is one of the key elements of Russia's socio-territorial structure. Additionally, the "heartland" occupies a large portion of the country's territory, but is still considered to be "excluded" territory. It is isolated from power centers and wealth by poor roads, bad communication networks, insufficient informational resources, and so forth. Everything eventually slowly permeates the heartland: modern employment, improved living standards, and media diversity limited to RTR (Radio and TV of Russia) and Channel One TV.

The country's size is usually considered a resource by politicians, but this is also the fundamental reason that so many people and even entire regions are excluded from modernization processes. We believe that only after the degree of spatial accessibility of basic social resources considerably grows, it will be possible to abandon the usual accents associated with poverty and equality. Many elderly people in the provinces do not have access to any resources, even in order to get to a hospital or social service facility.

Thus, opportunities for social exclusion are expanding. Instead of a specific person with his or her own biography or "life story," categories emerge – more precisely, "marginal population categories" – like the sick, elderly, poor, and alcoholic, etc. The process of this bureaucratic processing of a specific person with the technologies of modern social work, turning him into a member of a group that is either entitled or not entitled to aid, as N. Luhmann has comprehensively analyzed (Luhmann 2008).

Obviously, the bigger and more unlimited a government's power, the greater the role of its word. Moreover, the state's word differs from the word in its usual sense not only in that, when uttered, its echo flies through the mass media channels, but also as it becomes fixed in decrees and regulations that are soon turned into laws. In turn, these documents represent the painted map of social space ahead, where it will fit or rather be squeezed into Russian society.

In this space, social inclusion/exclusion is added to the visible line of poverty/wealth, which affects the elderly to various degrees. To what degree? We will investigate further in the upcoming chapters of this book.

Today's interest in exclusion problems as a result of ineffective socialization at any age is associated with the problem's significance in destabilizing Russian society. There are various approaches to analyzing ineffective socialization, for example: deviant career formation, social exclusion, and disqualification. The theoretical provisions of the concept of deviant career are based on the provisions extended by E. Goffman (Goffman 1959, 1971). A deviant career is obviously an atypical development scenario. We will pause on this description, which is based on data established in literature about teenagers forming deviant careers.

A person becomes a deviant by slowly getting used to (slipping in to) a socially condemned lifestyle. Moreover, so-called "normal" people have a very direct relationship to forming a deviant career, as these people in particular force others (teenagers, adults, invalids, and the elderly) toward the chasm through stigmatization ("branding," "labeling"), separating the stigmatized from those who have not been branded. This exclusion scenario passes through several stages or phases in its formation.

For an elderly person, for example, the first stage begins with the simplest things: family conflicts, lack or weakening of psychological contact between the older person and his/her adult children, dismissal from work, random arguments at the post office, in public transit, or at the health clinic, and so forth. The immediate environment, ambivalent, cold, and sometimes even cruel, gives them the first push toward exclusion development.

The next phase of the exclusion formation scenario is lifestyle restructuring from active ("overcoming"), to passive ("relaxing"). Expanding acquaintance with other people in these relaxing/outcast lifestyles gives a person his or her own sort of refuge from a wider social environment. Sometimes communities of "babushkas on benches" become a social niche that facilitates survival. Likewise, the influence of these interaction partners is substantial because it creates identification examples, helps to interpret situations, and so forth.

The complex internal process begins developing in the third stage, receiving the moniker "identity crisis." When it becomes known to those in one's close environment (relatives, neighbors, friends, and so forth), they can provide assistance. Depending on who is closer and more attentive to the person in crisis in a given situation, the exclusion process can here either be interrupted or continue development.

In the fourth stage, the person accepts his or her exclusion, rebuilds his or her world value, identifies with the elderly subculture, and in short: comes to terms with the social role designated by his or her stigma. Now, the "young and healthy majority" will accompany him/her to all bureaucratic institutions, including medical and social facilities, where a career is formed as an object of social aid, an outcast, and a member of the weak, etc.

The process of social exclusion, in principle, can be interrupted on any developmental level. However, the following two conditions are necessary for this to happen:

(a) The efforts undertaken by members of the "prosperous, normal" world much be directed at activating and respecting the elderly person's social needs;
(b) The elderly person must be ready to accept help from the "prosperous" world.

At the same time, it is important to underscore that the problems elderly people often encounter are in no way problems of "old age," although definitions to this regard may exist in official documents, which are entirely tolerant in idea and style: "There are specific problems inherent to old age: worsening health condition, decreased self-care ability, "pre-retirement unemployment" and decreased competitiveness in the labor market, unstable financial position, and accustomed social status loss. Elderly women find themselves in a particularly unfavorable position, indicative of long-term disproportion between the male and female population. A substantial portion of the elderly population is composed of migrants and people lacking residence and employment. Social and economic costs are growing for families caring for elderly relatives, and the family's faith in the quality of support sources for elderly people is decreasing. Solitary seniors and elderly married couples frequently end up in this unfavorable situation" (Concept project 1997).

Similar cases of "ascribing" certain problems to a particular age group encourage inadequate analysis of senior citizens' social position, while only politicizing its discussion. Isn't it probable that costs for families with children will rise, especially for those with handicapped children? Isn't it probable that orphaned children (solitary children) or solitary adults (after divorce or a loved one's death) will find themselves in an inauspicious situation? Yet, pre-retirement unemployment is considered symmetrical to youthful unemployment, because specialists say that both those who are "entering" and those who are "exiting" the workforce fall into risk groups related to employment. As a result, the cited excerpt from an official document is, in essence, stigmatizing.

***

From a sociologist's point of view, the range of problems affecting old age and youth are significantly connected with the way in which society tries to comprehend and resolve regularly repeating but at constantly changing and "updated" issues of its own propagation. The modern view of man, the idea of equal rights and every person's uniqueness dictates the notion of the absolute value of human life regardless of whose life it is – infant or elder, man or woman. The ongoing process of generational change requires that we not just look simultaneously at all of life's stages (birth, life, aging, death) retrospectively and prospectively, but that we also find meaning and necessity in each of them.

Still, the question resounds fairly often: "Is striving to live longer appropriate?" The question is related, in our opinion, to the fact that people who exhaust their life goals and do not know what to do with their old age, usually get very sick, age poorly, and even come to despise themselves.

References

Abrahamson, P. 2001. Sotsialnaya eksclusia i bednost. *Obschetstvennye nauki i sovremennost* 2: 158–166.

Andreev, E., and A. Vishnevskij. 2014. Skolko s soshkoj, a skolko s lozhkoj? *Demoskop Weekly*: 601–602. http://demoscope.ru/weekly/2014/0601/tema03.php. Accessed 6 Dec 2017.

Anisimov, V.N. 2008. Molekulyarnye i fiziologicheskie mekhanizmy stareniya. SPb.: Nauka. Srednyaya prodolzhitelnost zhizni v Rossii i stranah mira v 2014 godu. URL: http://bs-life.ru/makroekonomika/prodolzitelnost-zizni2013.html. Accessed 6 Dec 2017.

Bauman, Z. 2004. *Wasted lives. Modernity and its outcasts*. Cambridge: Polity.

Bourdieu, P. 2005. Razlichenie: socialnaya kritika suzhdeniya. *Ekonomicheskaya sociologiya* 6 (3): 25–48.

Concept project. 1997. (Proekt Koncepcii «O politike stareniya v RF» byl razrabotan vo ispolnenie punkta 1 perechnya meropriyatij federal'noj celevoj programmy «Starshee pokolenie», utverzhdennoj postanovleniem Pravitel'stva Rossijskoj Federacii ot 28 avgusta 1997 g. No 1090 «O federal'noj celevoj programme «Starshee pokolenie» na 1997–1999 gody». Period realizacii Koncepcii prodlen do 2010 g.

Demograficheskij prognoz do 2030 goda. 2013. http://www.consultant.ru/document/cons_doc_LAW_144190/b1dbbe242ed3e0f07fa8eeab64fa827ce6dc5b5c/. Accessed 6 Dec 2017.

European Commission. 2005. *Confronting demographic change: A new solidarity between generations. European Commission's green paper*. Brussels: European Commission http://ec.europa.eu/employment_social/social_situation/responses/a8118_en.pdf. Accessed 6 Dec 2017.

Eurostat, EUROPOP. 2013. *Population projections*. http://ec.europa.eu/eurostat/web/population-demography-migration-projections/population-projections-data. Accessed 6 Dec 2017.

Federal'nyj zakon. 2017. *"O ezhemesyachnyh vyplatah sem'yam, imeyushchim detej" (On every month benefits for the Family's First child) № 418-FZ. 28 dekabrya*. Accessed 31 Dec 2017.

Gilleard, P., and P. Higgs. 2014. *Ageing, corporeality and embodiment*. London: Anthem Press.

Goffman, E. 1959. *The presentation of self in everyday life*. Edinburgh: University of Edinburgh Social Sciences Research Centre.

Grigoryeva, I.A. 2006. Socialnaya politika i pozhiloe naselenie v sovremennoj Rossii: vyzovy i vozmozhnosti. *Mir Rossii* 1: 29–49.

Konstantinovskij, D.L. 2008. *Neravenstvo i obrazovanie: Opyt sociologicheskih issledovanij zhiznennogo starta rossijskoj molodezhi (1960-e gody – nachalo 2000-h)*. Moscow: CSP. http://www.socioprognoz.ru/index.php?page_id=85&id=18&param=http://www.socioprognoz.ru/files/File/publ/Konstantinovsky_Neravenstvo_Obrazovania.pdf. Accessed 6 Dec 2017.

Luhmann, N. 2008. *Law and states of exception*. Stuttgart: Lucius & Lucius.

Madrid international plan of action on ageing. 2002. Second World Assemble on Ageing, Madrid, Spain, 8–12 April 2002. https://www.un.org/development/desa/ageing/madrid-plan-of-action-and-its-implementation.html. Accessed 6 Dec 2017.

Maleva T., and A. Tindic. 2014. Na kachelyah polovozrastnoj struktury naseleniya (On the swings in the demographic structure of the population). Demoscop weekly. 10–23 fevr. http://www.demoscope.ru/weekly/2014/0585/tema02.php. Accessed 29 Dec 2017.

Neugarten, B.L., and G.O. Hagestad. 1979. *Age & life course. Handbook ageing & social sciences*. New York: Van Nostrand Reinhold.

Principy Organizacii obiedinennyh nacij v otnoshenii pozhilyh lyudej. Sdelat polnokrovnoj zhizn lits preklonnogo vozrasta. 1991. *Rezolyuciya 46/91 Generalnoj Assamblei OON ot 16 dekabrya 1991 g.* (Implementation of the international plan of action on aging and related activities). *UN General Assembly – Forty Sixth Session.* https://documents-dds-ny.un.org/doc/RESOLUTION/GEN/NR0/581/79/IMG/NR058179.pdf?OpenElement. Accessed 6 Dec 2017.

Ravenstvo v sfere truda – velenie vremeni. Doklad generalnogo direktora. 2003. *Mezhdunarodnoe byuro truda. (International labour organisation. Declaration on fundamental principles and rights at work).* Geneva. http://www.ilo.org/declaration. Accessed 6 Dec 2017.

Safarova, G.L. 2008. *Starenie naseleniya Rossii: Sovremennoe sostoyanie, perspektivy, voprosy socialnoj politiki. Pozhiloj chelovek v sovremennom mire.* St. Petersburg.

Shershneva, E. L. 2014. Reformy Harca: Povorotnyj put v politike zanyatosti i blagosostoyaniya v Germanii. *Zhurnal sociologii i socialnoi antropologii* 3.

Sidorenko, A., and A. Zaidi. 2013. Active aging in CIS countries: Semantics, challenges and responces. In *Current gerontology and geriatrics research.*

Smelser, N. 1988. *Sociology.* New Jersey: Prentice Hall.

Standing, G. 2011. *The Precariat. The new dangerous class.* London\New York: Bloomsbury Academic.

Taylor, P. 2010. *The careless state: Wealth and Welfare in Britain Today.* London\New York: Bloomsbury Academic.

The European Union and Russia – Statistical comparison. 2007. Luxembourg: Office for official publications of the european communities, rosstat – eurostat.

The Global Age Ranking. N.Y. UNFPA and HelpAgeInternational, 2012. www.helpage.org. Accessed 26 Nov 2016.

Tihonova, N.E. 2002. Socialnaya eksklyuziya v rossijskom obshchestve. *Obshchestvennye nauki i sovremennost* 6: 4–21.

Touraine, A. 1991. Face à l'exclusion. *Esprit 141.*

Yelutina, M.E. 2002. *Gerontologicheskoe napravlenie v strukture chelovecheskogo bytiya.* Saratov: Izd-vo SGTU.

Chapter 3
Age, Work, and Retirement: Quality of Life

3.1 The Main Approaches to Defining the Boundaries of "Advanced" Age

Two significantly different approaches have developed in understanding the nature and, accordingly, the age criteria and limits of aging: functional and chronological. The first, functional approach reflected the centuries-old specifics of social old age as a process of increasing loss in terms of personal physical health, the capacity for work, as well as the accompanying capacity for material self-sufficiency. Subsequently, in connection with increased life expectancy in elderly and old age, it also denotes lost ability for daily self-care.

Social old age, as is widely accepted in both domestic and foreign science, indicates a person's inability to provide for him or herself due to advanced age and, accordingly, his or her transition into being someone else's dependent. This definition issues from a qualitative approach, excluding any formal age groups, i.e. those determined by chronological basis.

The second, chronological approach to understanding the nature of social old age was formed in the context of the industrial-developed twentieth century society and reflects the birth of a fundamentally new social institution: pension provision by old age immediately demanded the introduction of a strictly formalized threshold to determine what constitutes "old age." The objectivity of this requirement is apparent, but so is "steamrolling" people of varying levels of activeness, work capability, health, and so forth, by this formally introduced "age of old age."

In other words, the qualitatively-functional aspect of old age was completely lost in the chronological approach. This resulted in a high degree of social heterogeneity among elderly groups that have been divided based on this approach.

If we refer to different age classifications, it is impossible to determine a single viewpoint on the beginning of old age in total accordance with common sense and the scientific concepts of the heterochronology* of development. The age periodization used to divide human life into different stages based on biological, somewhat

© Springer Nature Switzerland AG 2019
I. Grigoryeva et al., *Elderly Population in Modern Russia*,
https://doi.org/10.1007/978-3-319-96619-9_3

social, and economic characteristics is fairly complex. The number of human developmental "phases" occurring after birth may vary from 3 to 25 depending on different approaches. The simplest and most universal classification consists of three main periods: childhood, maturity, and old age.

Pythagoras, the ancient Greek philosopher and mathematician distinguished four periods of human life, corresponding to the four seasons: formation – "spring" (0–20 years), youth – "summer" (20–40 years), prime of life – "fall" (40–60 years), withering – "winter" (60–80 years).

Thus, metric time (life years or the number of years a human has lived) is a fairly arbitrary indicator, as long as one is not talking about the number of years a man has lived. Even an organism's general biological age may not correspond to years lived. The weakness of age classifications consists of eliminating social determinants of age. Thus, in the majority of Western countries, retirement rights are legally set to begin start at 65 years even later, as "old age" is there considered to begin 10 years later than for Russian women.

Heterochronology is the incongruity of development time constituting any complex process. In human life, physiological maturity does not correspond with psychological and social maturity. In advanced age, different aspects of aging also do not take place at the same time.

In recent years, through the course of pension reforms in Europe and North America, the retirement age moves toward 67–69 years and should reach 70 by the year 2050. The legally-designated retirement age per country will – at least in 20–30 years – be accepted by society as natural old age. The relative boundary between maturity (capacity for work) and old age (inability to work) for the statistically average individual can be generally determined only through such retirement-related cultural-legal norms.

3.2   Age Periodization and Age Limits Accepted in Russia

In Russia, the following age periodization is reflected in legal norms, including those relating to work and pension support:

Periods – Range
Newborn – 1-10 days
Infancy – 10 days – 1 year
Early Childhood – 1-3 years
First Childhood – 4-7 years
Second Childhood – 8-12 years for men, 8–11 years for women.
Teenage Years – 13-16 years for men, 12–15 years for women.
Adolescence – 17-21 years for men, 16–20 years for women.
Maturity I – 22-35 years for men, 21–35 years for women.
Maturity II – 36-60 years for men, 36–55 years for women.
Advanced Age – 61-4 years for men, 56–74 years for women.

Old Age – 75-90 years for both men and women.

Long-living Persons – 90 years of age and older for both men and women.

The international term, "the third age" is used to identify and study the elderly, i.e. those over 55/60 years as determined by the Code of Labor Laws and Russian pension legislation. Conventionally "old" people, i.e. those over the age of 75, are designated as members of the "fourth age." Old age still rousts fear as the phase of illness and infirmity preceding death; death's approximation forces people to seek ways to prevent aging or to age more slowly and healthfully.

The desire to be as healthy as in youth leads to the exaggerated role of medicine in defining aging (Rogozin 2012). On the other hand, aging or old age also has "secondary benefits" along the lines of illness, which we will establish below. In many ways, this is largely due to the interpretation that both illness and old age are ways of evading social responsibilities (Parsons 1954).

This brings us to two important points. The Roman doctor Galen (1st-second centuries CE) championed the idea that health in old age qualitatively differs from health in any other age and represents a sort of intermediate state between health and disease. The stereotype of "health loss with age" persists in modern understandings of aging. Therefore, instead of studying the specifics of health or health resources in old age, old age is studied as a type of ailment.

At the end of the nineteenth century, the state pension system in Europe also evoked recognition that an age-related border exists as each person becomes incapable of labor due to the onset of old age. In late nineteenth century in Germany and later in other Western European countries, a state pension system developed that clearly structured the retirement process, vesting it within an age range that determined the time when a person becomes incapable of working, i.e. old. In protestant countries, the established old age was as a rule 65 years for both men and women, which was determined by requirements of an effective insurance system that could not resort to borrowing from the state budget for pension support during the settlement period of 13 years (the average survival period) after 35 years of insured work experience. The logic behind this pension system differed from the logic behind the medical system. Hence, according to the World Health Organization (WHO) classification, "old age" is currently considered to begin at 60 years. This border was established in 1948 and, strange as it may seem, has not been yet been reevaluated.

It is widely thought that while diseases are curable to a certain extent, aging "cannot be undone," spawning age fatalism and the "patientization of old age" (treating the elderly like "patients") (Smolkin 2007). The perception of life's schedule has radically changed since concept of a "retirement age" emerged. To reach retirement age, i.e. become incapable of working, means to be liberated from the necessity to work, in any case, in terms of employment. As is still written in modern retirement documents: "Old Age Pension."

In the Soviet pension system, the principle of oldness as the age when a person is no longer able to work was initially disregarded in 1956 for the sake of demonstrating of socialism's superiority. Established work seniority was also lower than in the West. Understanding early retirement and short work experience as a superiority given by the socialist government was superimposed on Russian Orthodox work ethic, which does not encourage excessive labor effort. "The Russian Orthodox work ethic is oriented not on the "formal rationality" of economics, nor on striving towards efficiency for the sake of efficiency or towards industrial development for the sake of industry, but is fixed on a goal beyond economics and industry. Russian Orthodox work ethic doesn't create (like Protestant ethic) a spiritual basis for creating a "bourgeois type" of man. Rather it protects against this" (Koval 1994).

In modern Russia, according to approximately 25% of the population as surveyed by the Russian Public Opinion Research Centre (VCIOM), old age begins at the 60–65 years, while 18% believe that a person becomes elderly at 70–74 years. 2% of respondents suggest that people should be considered old at 40–49 years, while 1% believes that old age begins at 90 years. Incidentally, 5% of survey-takers think that old age begins at a different time entirely and 2% associate this concept with retirement (Older Persons 2009). In short, the survey participants acknowledge, albeit latently, that the social relationship, in this case to old age, is ambiguously biologically and chronologically founded.

The chaotic transformations in recent decades have not motivated people to form life plans or reflexive biographies. For the modern person, it has become natural to think in terms of risks and risk prevention. However, uncertainty has become the main risk, from which it is nearly possible to defend oneself, particularly on the state level (Social Theories 2008). The absence of deliberate strategies for socio-economic development that are understandable to the public in modern Russia exacerbates the representation of aging as dangerous to society, despite the fact that in many ways aging only shift emphasis from current areas of public regulation to new ones.

Healthy and active aging cannot result in the same end as when aging was legitimized as a "difficult life situation." The right to social services arose no later than the right to a pension, both beginning at the age of 55/60. A later age could have been set, as social services was a new regulation area (Federal Law 1995a, b). Widespread "access to services" motivated a significant portion of the population to age "more quickly and sickly." This is fully understandable from the viewpoint of the opportunity to receive aid in difficult life conditions as a consequence of those conditions transforming.

Altogether, this resulted in a contradictory combination of motives where no one wants to age but a significant number of people want to retire as early as possible, which is clearly achieved as the average retirement age in Russia is approximately 53 years for women and 57 years for men (Maleva and Sinyavskaya 2008). Understandably, no pension system can withstand such a burden. In addition, a society that has more retirees than elderly people looks fairly strange....

Psychologists believe that any age is associated with the expectation and fulfillment of turning points or crises (Erikson and Erikson 1998). However, this is just an abstract life cycle model – we believe that the interdependence of different life periods is far more important, i.e. when separate measures of life's trajectory in different situations and dependent on age can have different impacts on life's proceedings.

A more modern approach from the author's presentation "On Developing Human Potential" replaces standardized age "crises" with the concept of vulnerability or vulnerabilities. In the 2014 presentation, the change of vulnerability types over a lifetime are also considered in terms of the lifecycle. In contrast to statistical models, this analysis permits the conclusion that children, teenagers, and elderly citizens encounter different combinations of risks that in turn require targeted assistance.

Several life periods are particularly important. For example, the first thousand days of a baby's life, the transition from school to work, or terminating work life due to age. Complications during these life points might prove especially formidable and have long-term consequences. An entire list of such troubles may be ascribed to old age. These problems are, chiefly, waiting to retire, losing loved ones, and deteriorating health and financial situation.

The German sociologist Hans-Peter Blossfeld referred to these transitional states as passages, and almost all of spheres of a person's life can become unstable. The most important of these passages are education, work experience and the labor market, partnership, marriage and raising a family, and the later phases of life and age (Blossfeld and Huinink 2001). Naturally, the later phases of life are the most fragile.

The case may be that prognoses results will self-fulfill as people become older and program themselves to given events and believe that gradual deterioration of health and life quality are inevitable. These expectations are conditioned by objective reasons as well as prejudices and stereotypes in relation to old age. The elderly person's world is perceived as just a shadow of the past, "normal" world. However, elderly people should cease to be seen as just "old adults," and acquire a new quality.

Accordingly, Elaine Cumming and William E. Henry's once popular "liberation theory" subsides into the past, where it postulates that in terms of age hierarchy elderly people are distanced from younger social groups as a result of liberation from typical socio-professional roles – essentially, desocialization.

In accordance with A. Rose's theory on "subculture", having a firm affiliation to a social group of the same age helps aging people maintain their identity and sense of psychological stability (Smelser 1988). A.L. Thornton and S.A. Lishaev define old age as the life stage when human desire for higher meaning and values is fully realized, enabling ability to see and assess current reality anew.

Thus, a person does not just age biologically, he or she acquires a set of desired and expected behavioral patterns, experiences age socialization, and tries out new roles such as retiree, householder, grandparent, and so forth. However, with what an individual fills his or her "final" years depends entirely on that person and his or her own identity.

3.3 Increasing Elderly Employment

Population aging has already become one of modern society's critical problems, demanding social policy transformation. Meanwhile, there is currently no social experience anywhere in the world associated with life in an aging society, and thinking in terms of risks and threats is very widespread. Therefore, we believe it is necessary to change society's impression that an elderly person is absolutely bound to sickness, weakness, and malaise.

Judging from available publications, rejecting the concept of aging as synonymous with inevitable extinction and cascading ability loss proceeds much faster in the West than in Russia. This is related to traditions adopted at the end of the nineteenth century of a higher retirement age, as by the end of the nineteenth century and beginning of the twentieth century, elderly people were considered to be those over 65 years old. It is also connected to influence first of Protestant work sacralisation, and then of the consumer sacralisation. After all, elderly people born in the post-war years (baby boomers) constitute a large consumer group with the highest purchasing power. This is noticeable even in Russia, but public discourse is stuck on the idea that elderly citizens are "disadvantaged, ill, and in need of constant care."

The opinion that aging is a demographic and economic disaster, from our point of view, is just an effect of the usual horror stories stirred up by the media. The working population will not be able to "feed" the elderly only if technology does not develop and labor productivity doesn't grow, but that has not yet happened in the history of developed countries. However, the question remains more complex for developing countries or Russia.

Elderly Employment to Prevent Social Exclusion Before turning to elderly social exclusion analysis, we recall that "social exclusion/inclusion" is a relatively new theoretical framework, which suggests using certain concepts and schemes to replace the hierarchically-arranged "poverty/wealth" framework.

Furthermore, Russian sociology's conceptual apparatus has become more receptive towards the "non-class approach" and new concepts have come into use. Out from the shadow of class dilemma, groups have emerged that previously were simply considered "renegades" and were more often punished than attracted researchers' attention (homeless people, vagrants, prostitutes, and drug users, etc.). In modern anthropocentric sociology, reflecting on risks such as society's modernization and updating the "life scenario" or "scenario of an individual's life trajectory," and "aging risks," occupy a far greater place than before, and sociologists are actively developing language that is tolerant toward social abnormality.

Many researchers today note that, even in developed countries, the elderly are excluded from society and pushed to its margins. Relative deprivation related to retirement frustrates the symbolic order (Sztompka 1993).

The "social exclusion" concept is no longer new to Russian sociology. In recent Russian publications, we note a review of theoretical approaches to this concept (Dmitrieva 2012), a review of the possibilities of measuring elderly social exclusion (Saponov and Smolkin 2012), studies of elderly adaptation practices including lack of or low skill level (Rogozin 2012; Shmerlina 2013), and determining elderly inclusion in modern informational space (from basic tech usage skills (computers, cellphones, etc.) to Internet work skills (searching for information, making online purchases, accessing virtual educational resources) (Grigorieva and Chernyshova 2009).

Within the frameworks of a survey conducted amongst the Ivanovsk area's elderly people (2012), D. Saponov and A. Smolkin distinguished the following measures of elderly social exclusion: family and close relatives (their number, mutual aid level, and socialization intensity); friends and neighbors; work (continuing work after retirement, motivation to work and learn); and social activeness (intensity of communication and physical activeness, hobbies, computer use) (Saponov and Smolkin 2012).

It is fairly difficult to evaluate how Russia's pension reform situation influences elderly social exclusion, as the pension system is adapted to changing economic and demographic realities. The 2001 Russian pension reform ended with the decline of the personified accounting system, blurring the rules of pension indexation and increasing the arbitrariness of the current indexing solutions, which the reform's architect, renowned economist M. Dmitriev, announced long ago (Dmitriev 2005).

In a country where there is a significant sector of "grey-area" employment, listing pensions in accordance with official employers would leave a large portion of people pension-less upon reaching retirement age. In 2008, a way was selected to protect these people's interests, and the labor pension for a person holding a position for over five years little differs from the pensions of those who held a position for over 20 years. The state protected the rights of people in the zone of "grey-area employment" and discriminated against honest taxpayers.

This, in our opinion, became yet another source of violated trust between the government, employers, and employees, in consequence of which people themselves rush towards the moment of leaving the working community for the retirement community.

It turns out that the pension system dilemma was reformulated from "receiving a decent pension" to the dilemma of "the most minimal time and labor contribution to receive a pension." A portion of the population accomplishes this strategy by seeking jobs that give various seniority with right to an earlier retirement or seeking opportunities to form invalidity, which also provides the legal opportunity not to work and to receive a small monetary benefit.

Another portion of the elderly population perceives retirement as losing social status, although many say that they will continue working because it is impossible survive on a pension alone. A contradictory situation has developed: earlier surveys showed that 80% of the population did not want the retirement age to increase because retirement is perceived as a fully-deserved additional monetary benefit upon reaching 55–60 years of age. But now the opinion is changing: "The majority of working-age Russians do not support the proposal to increase required work experience to 35 years to receive a full pension. However, the majority approves of the idea to raise the threshold entitling the right to a pension from 5 to 15 years" (Public Opinion Fund (FOM) survey 2012).

Many people also do not want to retire within the first 5 years, that is, before 60–65 years of age. "If earlier the majority of people sought to stop working, now there is a growing number of people who do not want to retire, as well as those for whom the question of retiring isn't even plausible. The first group includes government employees, judges, professors, and scholars, who constantly fight for the right to work outside of the established limit. Under their pressure, changes are periodically made to legislation. There are growing ranks of people in liberal professions who are working as long as they can allow themselves" (Maleva and Mau 2013).

Thus, the situation connected with evaluating retirement prospects and elderly employment from the viewpoint of influence factors of continued/ceased employment on their social exclusion in Russia today is fairly contradictory and requires further study. In the case of continuing employment through the ages of 65–67, the question of exclusion significantly lessens.

Age Discrimination and Elderly Employment Elderly social groups are visibly heterogeneous. The process and pace of aging have an individual character, but in the present legal state the boundary used to define older age as the time when the right to "old age pension" vests was determined long ago. Traditionally, this age was 65, but in Germany under Bismarck retirement started at 70 although average life expectancy did not exceed 45 years due to high childhood mortality rates. In other countries, different pension schemes have been established, similar official retirement ages, and different actual retirement ages, as reflected in the historical-cultural and economic particularities in each location.

Today it is impossible to establish direct dependence between retirement age and expected lifespan in developed countries. The relationship between retirement age and the elderly population's proportion cannot be traced (Zakharov 1997). Thus, in Russia, the retirement age is low but there are significantly fewer elderly people than in developed countries, for example Sweden, where 65 is currently the retirement age for both men and women.

In the USSR, the pension age was supposed to demonstrate socialism's superiority, which is why it is different than in the West: 55 years for women and 60 years for men. In Russia, authorities simply feared changing the norm, although it has already changed in all other post-socialist countries, including Belarus. In addition, in recent years 35–40% of people annually receive a pension prematurely through preferential regimes.

This results in a large group of "young retirees" who are fully active and capable of working. "The proportion of persons above working age in the total population in early 2009 was 21.2%, while the proportion of retirees in the population was already 27.2% As a result, there were 562 retirees out of 1000 employees in the year 2008.

Moreover, we must consider that the relationship between insured people, for whom employers make contributions to the Pension Fund of Russia, and retirees, is still worse. This is because some of the employees work under the conditions of oral recruitment and informal wage agreements for small businesses that use the simplified taxation system, or are employed by sole proprietors, etc." (Maleva and Sinyavskaya 2011).

According to data from the second wave of socio-demographic survey, "Parents and Children, Men and Women in a Family and Society" (2007), in the first year after reaching retirement age, over 60% of retirees continue to work; approximately half of men and three fourths of women (Maleva and Sinyavskaya 2008).

Nevertheless, pensioners' occupation level remains fairly high over the first 5–10 years after retiring and then sharply declines.

The problem of senior citizens' employment is extremely relevant in modern Russian conditions. According to a prognosis by the RF Ministry of Labor, the number of retirees in Russia by 2020 will reach 39.4 million people (RIA Novosti 2013), and according to several experts' evaluations, by 2035 the number of working Russian citizens will approximate the number of retirees (RIA Novosti 2015). Government and legislative authorities are taking certain steps to find the optimal balance between elderly employment, options for retirement, as well as combining work and receiving social benefits.

In particular, in 2013 a law was enacted (Federal Law 2013), forbidding use in hiring notices from indicating the maximal acceptable age of job candidates. At the beginning of 2014, the Deputy Prime Minister, Olga Golodets announced, "In our country today, 12 million retirees continue to work. This is the most powerful proof that today in Russia there is a model of continuing work beyond retirement age, and women 56–57 years, and men 62–63 years all over the world feel themselves to be younger" (Golodec 2014).

However, thing in life are not always as they seem. At the Second National Conference on Aging held in Moscow from October 8–9, 2014, participants discussed the question of elderly employment and the perception of age restrictions by both workers and employers. Hence, the data obtained that for employers, the "older age group" is identified in the following way: beginning with retirement (55–60), but for individual companies from 35–40 or from 45 years.

The positions for older persons in all spheres are mostly technical: hosts, drivers, security guards, accounting, personnel departments, insurance agents, financial sector cashiers, industry workers, trade merchants, and nurses. It is noted that retirement has been a tangible loss, since there is a shortage of "cheap" workers, specialists (in health care), versatile workers (industry), etc. It takes time before new employees can fully replace the departing.

Employers, according to the afore-mentioned study and several others, adequately differentiate the "pros" and "cons" of having elderly employees. The "pros" consist of the following: experience and life wisdom; knowledge and qualification; positive influence of Soviet upbringing – honesty, responsibility, neatness, and duty; they never refuse (for example, to stay after working hours or leave for the weekend); they are well-founded and stable (they need to feed their family, they aren't "antsy" to move on to the "next thing"); they no longer have the problems associated with having small children; they can be mentors to younger workers; they aren't demanding in terms of salary; and they gladly participate in social activities.

The "cons" category includes the following: poor health; difficulty using computer technology; older people tire faster and are less capable of work; they are less creative and have a difficult time accepting anything new, they are conservative; they have less initiative and are less active; they display negative emotions and behavior; inadequate reactions; they frequently display symptoms of professional burnout; their knowledge has become antiquated and they have a narrow worldview; and they struggle to join a new community, etc.

In several studies on the motivation for continuing work, employers mention several conditions for keeping valuable elderly employees: decreased workload, flexible schedule (with salary decrease), the ability to take time off, and increased salary based on experience. It is puzzling that neither employers nor retirees have voiced the reason for continuing work in relation to recalculating the insurance part of the pension. Noticeably, the majority of retirees, unfortunately, do not differentiate recalculating base pension by the inflation rate and recalculating the insurance part of those working. However, the government has decided to end this recalculation "because of budget problems," although the workers themselves earn money towards this allocation.

After 2016, the size of the insurance part of the pension was recalculated, but the payment takes place for all years at once only after retirement. Understandably, this does not motivate work, at least in terms of official employment. It is worth speculating that the government will gradually refuse to recalculate the insurance part of the pension or will sharply decrease the size of the added amount.

The question about pension size corresponding to the living minimum is still not fully resolved, but now in all subjects of the Russian Federation retirees receive additional payments to meet the minimum wage if their pension is insufficient. In this case, the retiree is assigned to a social supplement from the Federal budget. The size of the supplement is determined thusly to complement a citizen's material security to meet the subsistence minimum.

According to data obtained in D. Rogozin's research, the age of 70 is considered the turning point, after which there is a sharp decline in proportion of working retirees (up to 14%). The main reasons for ceased labor activity after 70 are worsening health condition and decreased importance of work to older people (Rogozin 2012). However, even modern scientific approaches vary in terms of determining the limits of work capacity. According to international tradition, economically active ages are considered those within the period of 15 to 72 years, while Russian demographers believe that people older than 65 years are elderly (Andreev and Vishnevskij 2014).

When talking about economic motivation for employment, it appears to have grown not only in countries where retirees' quality of life has decreased as a result of reform, i.e. post-socialist countries, but also in countries that are fairly wealthy, for example, the USA. Articles have long been published about how American retirees use their basements to maintains family-owned/small businesses and are quite successful in this business, as well as in volunteering aid to the disabled and extremely old. This activeness was not originally associated with solely economic motivations or decreased quality of life, which have recently been observed. Additionally, in the USA, there has been a gradual transition toward retirement at 67 years, when the age had previously been set at 65 for many years.

True, senior citizens who have income in addition to pensions or property, in both the USA and in Europe, can retire early and premature retirement plans do exist for such cases. In the USA, the proportion of people working over 65 years in the year 2000 was 16.9%, in Japan it was 35.5%, while in Europe it is traditionally lower. Neverthless, the situation is changing: in recent years the European community adopted a decision to increase the retirement age to 67 years, and by the year 2050 to 69–70 years.

Answering the question of "why," we reiterate that elderly people work not just for economic reasons. For many, it is no less important to maintain the social status established in earlier adulthood, which secures respect from others as well as the level of earned payment. In modern Russia, professional success competes with family values even amongst women, and accordingly it is difficult refuse these social and socio-psychological values, solely because of reaching a certain age.

As a result, for elderly people whose interest in work was sustaining, connected with self-actualization and positive sociality, retirement is far more complicated than for someone who has simply earned a living. In addition, people in tedious or hard labor occupations frequently assert that they cannot wait to finally retire, which is entirely understandable.

Guaranteed employment for senior citizens is also needed because it is a motivational dominant that is aimed at maintaining successful work experience and social acknowledgement, which can even grow in old age. Because of this, social service for the elderly should include professional orientation and preparation in order to seek decent work for people in the pre-retirement age and for the elderly, especially in the first decade after reaching retirement age. This will encourage slower aging rate and preserve working potential, as well as renewing senior citizens' work capacity level. This is particularly important in Russia, where the current employment situation discriminates toward the elderly and is constructed on retirement norms that were established in 1930–1950.

At that time, the technology level, working conditions, and, accordingly, rate of age-based work capacity loss were completely different, and no one had yet even suspected the possibility of an population aging crisis. As a result, elderly employment today pertains to the question of changing society settings and elderly people's motivations, which should be supported by legislature and employers. This brings us to yet another classic gerontology work by Alex Comfort who stated that "any increase in life expectancy, created by slowing the aging process, should represent a period of increased labor experience for societal members…" (Comfort 1964: 36). The political decisions associated with employment and elderly social status also influence the way that society views these problems, and which are highlighted.

There are also many people who create a "second career" and become occupied with social work, becoming volunteers and activists in social organizations. The problem, in our opinion, is that their activity doesn't receive due acknowledgement from society and is often instead perceived as lunacy. Thus, in the Concept of Developing Voluntiring in Saint Petersburg, adopted in 2008, there was emphasis on increasing social activity amongst the youth, but the elderly population was again disregarded.

In a series of brochures on voluntaring development issued by the philanthropic society "Nevsky Angel," elderly people are also considered the recipients, not agents of volunteering. According to data collected by T. Smirnova, "16% working retirees and 5% non-working indicated that social activity is a means of combating loneliness. Meanwhile, according to 69% of employed retirees, it is continuing work in particular that helps fight loneliness" (Smirnova 2007). However, in recent times, "silver volunteering" (i.e. elderly) is rapidly developing.

Accordingly, there is need not only for short-term campaigns like "Thank you, grandpa, for the victory," but also for acknowledging the important contributions elderly people have made in common daily life, for example through neighborly mutual aid and by supporting lonely and ailing elderly people, etc. The presence of this a support network is better seen by Western researchers (Keating and Dosman 2009; Harris 2011), as we are more accustomed to studying aid provided by social services, and natural forms of elderly sociality are still infrequently studied by Russian authors, with the exception of the works of N. Shchukina (Shchukina 2004).

On the one hand, developing volunteer positions as well as preserving part-time occupations can be considered the most important prevention tool in terms of elderly social exclusion, and on the other hand is a means to fill a number of jobs in the system of serving the population that are unattractive to young people. If the reason for elderly workers' difficulty adapting is truly connected to age, then steps need to be taken to improve work conditions and production environment, as well as reducing the duration of the work day or work week, transitioning elderly workers with piecework into the system of per-time-pay, but not discharging them.

3.4   Representation of Elderly Employment Opportunities in Information Space of the Internet

Co-authored by A.V. Dmitrieva and L. A. Vidiasova.

In preparation for conducting qualitative empirical research with regard to the problem of elderly people in the labor market and opportunities for their inclusion in social life and paid employment, we attempted to examine this context in cyberspace. We studied several specialized sites (N = 32) with postings about work for elderly people, data from the informational bulletin Yandex.Rabota from July 28, 2011 (later material from Yandex is more specialized-sectoral in nature), as well as updated data from the services Yandex.Rabota and Trud.com in the autumn of 2015. The service Yandex.Rabota is a vacancy aggregator, collecting notices from over 100 sites, including HeadHunter.ru, SuperJob,.ru, Rabota.Mail.ru, Rabota.ru, and Job.ru. The service Trud.com accumulates vacancies from various job search sites.

In June of 2011, more than half a million actual work offers from over one hundred thousand employers were posted on Yandex.Rabota (Rynok truda 2011). We were interested to find out which vacancies in particular are offered, as well as the presence discriminating, i.e. ageist, notes. Consequently, the main research question at hand can be formulated as follows: how are employment opportunities for the elderly represented online and are these representations discriminatory in nature?

Names of Sites Offering Vacancies for Elderly Job-Seekers The first thing that caught our attention and even surprised us when surveying specialized sites is how they are named. The majority of sites offering vacancies for elderly feature the category "pensioner/retiree" in their names. This category indicates the formal status of a person exceeding the age fixed by legislature as the beginning of retirement. Meanwhile, the separate category of "senior citizens" seems completely unused. As shown above, in Russia it is possible to be retired without being elderly, but rhetorically this difference is not traced. Accordingly, Internet users employing these resources have no doubts about the interchangeability of these concepts. It seems that either this differentiation does not exist for them, or they are certain that that these concepts are interchangeable.

A portion of specialized sites possess an enlightening-educational or recreational nature, while some offer only a small quantity of vacancies. The main portion of sites offer nearly identical vacancies, which are limited to housework or other low-skilled labor. However, a year ago, on a site with the discriminating name "Baba-Deda" (i.e. Grandma-Grandpa) there was a posting for the exotic vacancy as a "Feng-Shui garden" specialist and the more traditional insurance agent, while this year there are also two vacancies: kiosk worker and call center representative (http://baba-deda.ru/offers/2212).

Vacancy Structure According to the informational bulletin Yandex.Rabota from July 28, 2011, there are candidate age limits per every third vacancy. Most often, the age limit is present in announcements regarding physically difficult work: mechanics, geologists, and construction workers. Age is listed more rarely in postings for work in the banking sphere, art, and culture fields. Of course, when discussing physically difficult work, ageism is not truly applicable, however limitations in the other afore-mentioned situations are surprising. Thus, elderly people are most excluded from the work in the following fields: trade, marketing, and advertising. Moreover, it seems that a large portion of the adult population in general is also excluded in this situation, as in 40% of vacancy postings in these employment spheres seek young people under the age of 25. In addition, people of over 45 years of age are most frequently needed as housekeeping helpers and nannies; more than 40% of all postings in this field have a minimum age requirement. Furthermore, housekeeping or nanny work doesn't seem to us to be explicitly physically easy, as it assumes the burdens associated with the need to take children by the hand or pick them up, keep up with them, and bend over in the process of cleaning, etc.

The results of analyzing vacancy portals in the year 2015 allowed us to evaluate the level of vacancy offerings for older-aged people. With the help of the Yandex. Rabota system search, we found 35 vacancies in which preference toward retirees was explicitly indicated. Interestingly, one third of these offers related to work from home on a computer (dispatcher, web designer, consultant, etc.). The majority of found vacancies offer part-time employment and promise retirees a salary from 6 to 30 thousand [rubles] a month. Aside from that, analogous to data from the year 2011, we found work offerings for retirees that require significant physical burdens: promoters, couriers, handymen, conductors, and guards, etc.

The service Trud.com allowed us to accumulate data about vacancies throughout the entire country and according to data from October of 2015 we observed 1.5 thousand vacancies accessible to Russian retirees. Just over half of these (54%) did not require work experience similar to the proposed post and approximately one tenth offered work for retired veterans. Predominantly, the pay level for these found vacancies (49%) did not exceed 20 thousand rubles per month.

More than half of the vacancies offered to retirees (54%) provided work with a changeable or flexible schedule, remote work regime, or a part-time work day.

Interestingly, sites with vacancies are addressed toward an elderly audience that, by all appearances, uses Internet resources, and the number of elderly users is notably growing, as confirmed by data from Yandex.

The Content of Several Publications Dedicated to Older Age Employment. In the contents of analyzed specialized sites, in addition to published vacancies for older people, there are also articles of informational-advisory nature.

Hence, the Russian-language site "Modern Internet-Journal for People 50+ with Active Lifestyle" (the site's author lives in Hannover, Germany), offers the following "Ten Ideas for your First Business – After 50 Years of Life" (Ten Ideas… 2011):

1. Develop content websites dedicated to various life problems, such as medical information, fraud and scams, investment information, and receiving commissions;
2. The perception of aging is changing: fitness organization for older people, active vacation and extreme vacations;
3. "Kindergarten" for senior citizens: daily care for elderly and people in places that are specially designed to accommodate this activity, home care for the ailing;
4. Employment service for older people who still want to work at least part-time;
5. Financial consultant for elderly clients, real estate and investment planning, financing long-term care or housing options;
6. Driver for elderly people who need transport around the city;
7. Housekeeping assistant: Preparing food, buying groceries, paying bills, and offering care;
8. "Simply Talk," read a book, help someone go for a walk;
9. Pet-sit while owners are out of town;
10. Publish journals or informational surveys for the older population.

The target audience for these ideas seems somewhat "blurred." On the one hand, the article is about "first business after 50," which resonates with the general name of the e-journal which is addressed to people over the age of 50. On the other hand, it arises the questions connected with the choice of this particular age when retirement is still 10–15 years away (as we recall, the text's author is from Germany, where retirement begins at 65). Furthermore, some of the ideas suggested requiring access to significant capital. For example, the author estimates the start-up costs for creating a content website to be 5–25 thousand dollars. Apart from this, some of the ideas suggested require fairly serious qualifications as well as substantial start-up capital (for example, financial consultant or magazine publisher). Finally, it is difficult to call pet-sitting, reading a book, or housekeeping "business ideas." These ideas partially overlap with offers from purely Russian sites. However, the more complex situations should be accompanied by at least minimal instruction, considering the actual diversity of the elderly audience.

Consequently, we draw the following conclusions. As a rule, the category "retiree/pensioner" features in naming specialized sites, while "senior citizen" does not. This essentially makes these two concepts equivalent. The structure of offered vacancies is uniform and includes occupations that are stereotypically assigned to the elderly, or in contrast, require too high qualifications and/or access to financial resources. Meanwhile, there is a growing number of older Internet users, i.e. they are included in modern informational space, which makes less opportunities for social and professional discrimination and expands available employment opportunities.

Elderly Employment Based on Studies of Social Media The informational space of social media presents an enormous layer of information for discussing different types of phenomena. The author's collective tasked itself with determining which questions concern Russian citizens in the context of elderly people's employment, their employment opportunities, and discrimination factors. The commercial instrument IQ Buzz (http://iqbuzz.ru/) was used to track the dynamics of all messages, making it possible to trace the appearance of messages on the topic and their dynamics by using a given selection of key words.

As a result of analysis over the period of January 1, 2010 through May 15, 2015, 5765 documents and social media messages were found on the given study topic, which have been classified in the following way: comments, forum messages, microblog statuses, videos, notes, news, and other posts.

During the study we noted increasing activeness in discussing the given theme and several informational bursts associated with relevant newsbreaks.

In particular, since January 2014, people actively discussed opportunities for volunteer work for senior citizens while organizing the Olympic Games in Sochi. Online, it was possible to find news stories, as well as the personal reports of individual retirees who partook in this activity. Thus, we must emphasize that such activities, even if not paid, are good examples of increased elderly inclusion in an active social life. Messages online also highlighted the sentiment that voluntarism lets retirees feel needed by society.

The most active discussions online in 2013–2014 related to advice publications about how to work from home, distributing working space and time, as well as options for earning money online. Site user interest (gauged by a large quantity of "likes" and message comments) attracted the notice of Harvard University psychologists. In June 2013, they proved that the aging brain absorbs a larger quantity of information, than younger people's brains, but does so more slowly and reliably, connecting it with other phenomena and classifying it. As an example of the practical use of this knowledge, they referred to major foreign companies that entice retirees to come back to work (for example, Walmart).

Apart from this, the Internet offers a place to discuss unique cases where retirees received work. In particular, in December 2014 one of Melbourne's homeless retirees, Natalie Trayling, settled in a park and quickly became a local attraction by playing a children's piano and sometimes stopping by a music store with the owner's permission to play real instruments. Ultimately, she was offered a job playing the piano in a hotel lobby, which she continues to do today.

Based on research results, we noticed that adopted legislative initiatives and government measures on this issue are practically immediately reflected in online discussions. For example, May 19–25 2014 was marked by a surge of discussions in the period leading up to the Ministry of Labor's annual presentation about the socioeconomic position of elderly people in Russia. On 10/10/2014, the First Deputy Minister of Labor and Social Protection of the Russian Federation Aleksey Vovchenko stated that the Russian Ministry of Labor planned to complete work on preparing the Strategy in the Interests of Older Persons by spring of the following year, prompting active discussion from the period of October 27 – November 2, 2014. On February 2, 2015, when the project structure of the "Strategy in the Interests of Older Persons" was published, a new wave of discussions and comments emerged, continuing through year, than the Strategy was adopted (in 2016).

In March–April 2015, people discussed the correction the Russian Federation's state program "Social Support for Citizens" as it was supplemented by the "Older Generation" subprogram. In 2015, these gradually increasing conversations were augmented by analogous discussions relating to the holiday May 9th, Victory Day, which traditionally attracts community attention to the various problems of veterans and other retirees.

We also noticed that messages about elderly s employment are published primarily on the social network VKontakte, but a substantial portion of messages also appear on various personal Live Journal blogs. In terms of message types, posts are the most popular (32.8%) and comments (23.5%). The most popular sources were the following: Vkontakte.ru, Livejournal, Liveinternet, Twitter, Goolge+, Mail.ru.

Based on information obtained using the service IQ Buzz, documents with a neutral informational tone dominate amidst messages on the topic of "Elderly Employment." This monitoring instrument allowed us to determine the sex and age of message authors, if they listed this information in their accounts. Distributing author's messages by gender sign did not expose any serious differences between men and women's interests on the given topic: the portion of female authors accounted for 56.5% of texts while the portion of male author's accounted for 43.5%. Amongst all authors, sex was automatically determined for approximately 47%.

From the beginning of 2011, the audience of discussed topics related to elderly employment as well as users on internet sites who view these discussions consisted of more than 26 million people. Citizens of ages 26 through 35 showed the greatest interest in the studied topic. The issues researched also roust interest from people of pre-retirement and of younger retirement age (age was determined for 18% of authors).

In addition to federally significant news or stories from international practice, people in different regions discuss regional strategies to facilitate employment for the population. In particular, in March of 2015, a by the Cabinet of Ministers of the Republic of Tatarstan published resolution No. 186 "On Amendments to the State Program, "Employment Assistance for the population of the Republic of Of Tatarstan from 2014-2020"."

The analysis conducted permits the following conclusions: the online community actively discusses vacancies eligible to elderly people, as well as various changes in labor conditions that are associated with retirement age (discussions encompassed 26 million people). As a result of this analysis, we conclude that the community has begun to take active interest in elderly people's employment and to continue discussions with ever growing interest, the greatest of which is displayed by people who are still fairly distant from retirement age.

Informational messages online reflect a surge in real interest toward the topic of elderly people's employment and most often describe:

- actions of authorities to form policy, strategies, and legislation in this field;
- possible work options for retirees;
- unique cases of elderly employment.

In recent years, the elderly employment level in Russia has grown, and the labor market currently offers a substantial number of jobs with flexible employment or similar regimes that suit people aged 70. Job offers have also risen for elderly people with key requirements including computer proficiency level. The motivation to continue working can be dictated by the positive desire be needed by others, to maintain one's social status, and to feel fully-valued as a member of society.

Employment helps adapt to social changes, while increasing inclusion in informational space lessens the qualification gap between generations. There are still more expansive opportunities for community work helping children, handicapped, and the extremely old, to which it is important to attract "younger" elderly people. Senior citizens have a demand in terms of transferring knowledge and experience to the young; many live with families or children, or maintain close relationships with them through mutual care.

***

Sociologists need to pay attention to life-building experiences of successful elderly people who have long since crossed the threshold of retirement age. These individuals are becoming increasingly common in science, education, and cultural activity, and their renown and career have quickly developed, particularly over the past 20–25 years, i.e. already in retirement age, despite the historical drama associated with this period. Macro-events of the transformation period gave the push towards intensive development on the micro-level for actively adapting personal own activities.

3.5 Conclusion

What are the opportunities for sociology in rethinking aging and old age? The category of age is now recognized as socially-constructed, suggesting the existence of various types of social experience maturing and aging. The most important identifier is employment and its nature, since continued employment is a means of life extension.

Here, sociologists can find support in the field of medicine. According to renowned gerontologist Vladimir N. Anisimov's estimates, maximal long-living persons are observed amongst artists, performers, and musicians, whose employment virtually has no predetermined age (Anisimov and Zharinov 2013).

Delaying non-superficial pension reforms gives us the opportunity to draw on the experience of others. Sweden and Great Britain are very in their different approaches to social policy, but are both quickly aging and they have already rejected the normally established "retirement age" and extended employment past the age of 65. Employment and self-employment provide high incomes to those who continue working after 65 years, though it is true that they increase the gender gap (Arber 2013). Therefore, lowering the level of unequal income for both elderly and younger people, we immediately encounter other, no less important, aspects of inequality.

This leads us to understanding that pensions/retirement age is an antiquated institution that leads to a growing population group's social exclusion. It is difficult to imagine that in modern conditions it is possible to engineer "confidence" for 30–35-40 years into the future. The answer could be to introduce a system of age payouts and individual or professional insurance starting from 60–65 years. Maybe then people will begin to plan out their lives earlier, and not just when retiring....

Russia's singularity is the leading role power-holders play in cultivating destructive socio-economic practices, which was evident in the studied issue. To this regard, instead of analyzing the reason for deviations from the intended results of the Pension Reform enacted in 2002, authorities continue to seek reasons for its failure in demographic processes and other very obvious things.

People forget that modern technology, growth in labor productivity, and carefully collected social deductions can radically change the relationships between employed and retired people – relationships that are needed for normal insurable retirement. Instead, a new round of radical changes to the pension system has begun, undermining people's remaining trust in the government.

"It seems," Marina Elutina and co-authors assert, "that studying the mechanisms that include people retiring due to age, in the context of a full life, the correlation between their views and existing social norms, and their adaptation to life's structural changes in connection with the transition to retiree status, significantly advances the development of a unified concept of state social policy in respect to this category of citizen. Sociological reflection on sources and factors of tension relief are imperative, the search for more effective ways, forms, means, and methods of achieving an acceptable living standard level during the transition to retiree status" (Elutina and others 2006).

The rapid changes characteristic of modern society give way to the feeling of losing one's place in history; people have trouble drawing the line between past, present, and future, reaching orderliness and integrity in planning life. The line between these stages is more accessible in a linear life flow, but, if life branches, it often does, then these lines become blurred or slipped through.

The media exploits the image of elderly people as wretched, miserable, and poor, reducing all of their problems to low pensions. Mass media channels immediately grasped at the topic that there is insufficient money to pay pensions due to the increased number of elderly, but never referenced existing tax underpayment or the of the pension system's actual failures, due to which reform is essential.

It is the senior community's best interest to highly well-being or misfortune for elderly people more objectively. For example, the "post-war baby boom generation" is fairly numerous everyone and currently underrated as active consumers. Their situation, on average, is far better than those born 10–15 years earlier. They are active participants in many "consumer society" practices, and moreover successfully and even unexpectedly turn the same practices using Information Communication Technology (ICT) in pure consumer practices, which we will discuss further in Chap. 5. We believe it is entirely plausible that in the near future elderly people will carry out travel/tickets orders with their discounts and other purchases online. In fact, this has been noted several times already by marketing experts.

Frequently elderly people are written about as the victims and as discrimination subjects. Victimization, however, is not a harmless discourse mechanism, as the "victim" always requires service or external care. We agree that the "state sacrifices subjects of their right in order to indicate power and the ability to regulate their behavior," (Kondakov 2015). Responsibility for violating elderly citizen's rights, for their position, nevertheless, is essentially left to the elderly themselves – "they are too numerous."

In terms of "age's liberation," it is essential to rethink the relationship of authority penetrating intergenerational relationships. Then, it is possible, that introducing changes in life schedule, at least for the educated part of society, will occur without a sense of systemic violence. This must be achieved so that pension relationships finally become understandable to the population and to lessen segregation by age.

In Russia's conditions, associated with high potential vulnerability, poverty, and social exclusion, it is extremely important to stimulate all types of social activity for different groups of elderly people and to track the dynamics of these indicators to develop effective measures for supporting the these groups' quality of life.

The state's maniacal desire to spend less on retirees, especially accompanied by massaging the problems of population aging in the media, looks strange in the budget, considering that since 2014 and there have been more expenses on defense than on retirement. Despite the particular popularity of such a "militant" policy amongst the population, it is necessary to spend more on education, health protection, and culture for the country's development, i.e. to boost human and not defensive potential.

The political task at hand is to change the image of old age, to reject pitying humanism in favor of more respectful and liberal relations, to normalize employment beyond the limits of retirement age, and to stop associating it with the extreme need, forcing the elderly to continue working. It is possible that this is also a basic question in terms of the country's socio-economic development or updating social policy.

As a result, for the time being it would be premature to conclude that Russia's elderly population is threatened by the risk of social exclusion associated with retirement age or losing work and qualification.

References

Andreev, E., and A. Vishnevskij. 2014. Skolko s soshkoj, a skolko s lozhkoj? *Demoskop Weekly*: 601–602. http://demoscope.ru/weekly/2014/0601/tema03.php. Accessed 6 Dec 2017.

Anisimov, V.N., and G.M. Zharinov. 2013. Prodolzhitel'nost'zhizni i dolgozhitel'stvo u predstavitelej tvorcheskih professij. *Uspehi gerontologii* 26 (3): 405–416.

Arber, S. 2013. Gender, marital status and intergenerational relations in a changing world. In *Global ageing in the 21st century: Challenges, opportunities and implications*, eds. by S. McDaniel and Z. Zimmer, 215–234. Farnham: Ashgate.

Blossfeld, H.-P., and J. Huinink. 2001. Lebensverlaufsforschung als sozialwissenschaftliche forschungsperspektive: themen, konzepte, methoden und problem. *BIOS* 2: 15.

Comfort, A. 1964. *Ageing: The biology of senescence*. London: Routledge & Kegan Paul.

Dmitriev, M.E. 2005. Novyj strategicheskij vybor: mnimye i real'nye ugrozy dlya pensionnoj sistemy. In *Laboratoriya pensionnoj reformy. Informacionno-analiticheskij portal* http://pensionreform.ru/42409. Accessed 6 Dec 2017.

Dmitrieva, A.V. 2012. Socialnoe vklyuchenie, isklyuchenie kak princip strukturacii sovremennogo obshchestva. *Sociologicheskij zhurnal* 2: 98–115.

Erikson, E., and Erikson, J. 1998. The life cycle completed: Extended Version. W. W. Norton.

Federal Law of the Russian Federation on social services for older persons and persons with disabilities (in Russian). 1995a. https://www.rg.ru/1995/08/04/socobslujivanie-dok.html.

Federal Law of the Russian Federation on the foundations of social services for the population in the Russian Federation (in Russian). 1995b. http://ivo.garant.ru/#/document/105642:0.

Federal Law of the Russian Federation on amendments to the law of the Russian Federation on employment in the Russian Federation and certain legislative acts of the Russian Federation N 162-FZ. 2013. http://www.consultant.ru/document/cons_doc_LAW_148473/.

Golodec, O. 2014. Tema povysheniya pensionnogo vozrasta zakryta na 10 let. *Finmarket* 17.01.2014 http://www.finmarket.ru/economics/article/3607804. Accessed 6 Dec 2017.

Grigoryeva, I.A., and S.P. Chernyshova. 2009. Novye podhody k profilaktike socialnogo isklyucheniya pozhilyh. *Zhurnal sociologii i social'noj antropologii* 2: 186–196.

Harris, J.G. 2011. Serving the elderly: Informal care networks and formal social services in St.-Petersburg. In *Gazing at welfare, gender and agency in post-socialist countries*, ed. Maija Jappinen, Meri Kulmala, and Aino Saarinen. Cambridge: Cambridge Scholars Publishing.

Keating, N., and D. Dosman. 2009. Social capital and the care networks of frail seniors. *Canadian Review of Sociology* 46 (4): 301–318. https://doi.org/10.1111/j.1755-618X.2009.01216.x.

Kondakov, A.A. 2015. Otrazhenie migracionnoj politiki v oficial'noj presse: subyekty v mediaprostranstve. In *Vestnik Sankt-Peterburgskogo gosudarstvennogo universiteta, Ser. 12.* Vyp, vol. 1, 147–154.

Koval, T. 1994. Etica truda pravoslaviya. *Mir Rossii* 2: 81–93.

Maleva, T., and V. Mau. 2013. *Chetyre strategii obespecheniya starosti.* http://www.vladimirmau. ru/ru/rss/item/readarticles/tatyana_maleva_vladimir_mau_chetyre_strategii_obespecheniya. Accessed 6 Dec 2017.

Maleva, T.M., and O.V. Sinyavskaya. 2008. Nuzhno li povyshat zanyatost pensionerov? In *DemoskopWeekly*, 341–342 http://demoscope.ru/weekly/2008/0341/tema04.php. Accessed 6 Dec 2017.

———. 2011. Povyshenie pensionnogo vozrasta: pro et contra. In *Obsuzhdenie «Strategii 2020»* http://2020strategy.ru/data/2011/07/14/1214719869/4.pdf. Accessed 6 Dec 2017.

Opros FOM: Stazh dlya polucheniya pensii. 2012. *Opros FOM*. http://fin.fom.ru/ekonomika/10682. Accessed 6 Dec 2017.

Parsons, T. 1954. Values, motives, and systems of action. In *Toward a general theory of action*, ed. T. Parsons and E. Shils. Cambridge: Harvard University Press.

RIA Novosti. 2013. *Pensionerov v Rossii v blizhajshie tri goda budet vse bolshe.* http://ria.ru/society/20130705/947941313.html. Accessed 6 Dec 2017.

———. 2015. Siluanov razyasnil, *zachem nado povysit pensionnyj vozrast v Rossii.* http://ria.ru/economy/20150206/1046365942.html. Accessed 6 Dec 2017.

Rogozin, D.M. 2012. Liberalizaciya stareniya, ili trud, znaniya i zdorov'e v starshem vozraste. *Sociologicheskij zhurnal* (4): 62–93.

Rynok truda i poisk raboty v internete. 2011. http://download.yandex.ru/company/Yandex_on_Jobsearch_2011.pdf. Accessed 6 Dec 2017.

Saponov, D.I., and A.A. Smolkin. 2012. Socialnaya ehksklyuziya pozhilyh: k razrabotke modeli izmereniya. *Monitoring obshchestvennogo mneniya* 9 (10): 83–94.

Shchukina, N.P. 2004. *Institut vzaimopomoshchi v sisteme socialnoj podderzhki pozhilyh lyudej.* Moscow.

Shmerlina, I. 2013. Liberalizaciya stareniya: teoreticheskie illyuzii i ehmpiricheskie anomalii. Monitoring obshchestvennogo mneniya: ehkonomicheskie i social'nye peremeny 3: 165–175; 4: 71–83.

Smelser, N. 1988. Sociology. N.J. : Prentice Hall.

Smirnova, T.V. 2007. Perspektivy zanyatosti pozhilyh v usloviyah demograficheskogo postareniya. *Zhurnal sociologii i socialnoj antropologii* 2: 123–133.

Smolkin, A.A. 2007. Medicinskij diskurs v konstruirovanii obraza starosti. *Zhurnal sociologii i socialnoj antropologii.* 10 (2): 134–141.

Social Theories of Risk and Uncertainty: An Introduction. 2008. Edited by Jens O. Zinn. Malden/Oxford/Carlton: Blackwell Publishing.

Sztompka, P. 1993. *The sociology of social change.* Oxford: Wiley-Blackwell.

Zakharov, S., and G. Rahmanova. 1997. Demograficheskij kontekst pensionnogo obespecheniya: istoriya i sovremennost. In *Sovremennye problemy pensionnoj sistemy: kommentarii ehkonomistov i demografov.* Moscow.

Chapter 4
What Does It Mean to Be Old? "Elderly" Identity as a Sociological Problem

Modern identity researchers pay special attention to transforming identification practices connected with value pluralism, the process of globalization, knowledge societies' formation, and other current processes.

This set of problems is interpreted from different perspectives: P. Bourdeau (Bourdeau 1998) questioned the social mechanisms assigning values to different identities, Anthony Giddens (Giddens 1990) determined the modern identity features of an active individual in the age of globalization. J. Habermas (Habermas 1984) described the formation of the "social state client" attacked by the "system"; Z. Bauman asserted that the identity of a person is individualized by the era and abides in a state of social uncertainty (Bauman 2004), and etc.

For a long time, age was considered to be an anthropological category, rather than a sociological one. As such, age-homogenous groups entered sociology gradually, beginning with Charlotte Bühler studying teens' journals, through young "groups on street corners" by Chicago sociologists, and finally becoming a sociological subject during the 1968 student riots. In addition, there was never any doubt that juvenology was a part of sociology. Meanwhile, elderly people were referred to medicine, social work, and psychiatry fields, so the distinguisher "social" has to be added to the term "gerontology" today to enter the sociological field.

As is well known, no social phenomenon can exist in isolation from the historic systems surrounding it. Hence, elderly people are not an abstract-ideological category and in recent years the understanding of elderly people, advanced age, or the age of growing old is being intensively reconsidered. We see a continuum of elderly situation assessments, from "the state deprived us" to understanding that the notable increase in the number of elderly people is an effect of slowed aging processes and increased possibilities for living longer, which in itself is happiness, to a certain extent, considering that a long life is naturally preferable to a short one.

© Springer Nature Switzerland AG 2019
I. Grigoryeva et al., *Elderly Population in Modern Russia*,
https://doi.org/10.1007/978-3-319-96619-9_4

4.1 The Gender Approach and Studying Elderly Identity

Considering the various directions of feminism (sometimes in the West people even refer to "feminisms"), in seeking analogies for understanding ageism, in the most general way one could say that the key topic is the dilemma of similarities and differences between men and women, accordingly, between the young and old. As professor Hester Eisenstein writes, "the topic of "differences" became integral for modern feminist thought, at least from the time of Simone de Beauvoir's publication of "The Second Sex," and in particular, from the time of the women's revival movement in the 1960s (Eisenstein and Jardine 1990: XVI).

J. Evans also notes that feminist schools are "structured around topics of equality, similarities, and differences" (Evans 1995: 160). Avoiding the question of similarities and differences between men and women is impossible, because all traditional culture is based on the principle of contradiction and hierarchy of male and female origins and, accordingly, from whence stems the stratification ideology of men and women in society. Particularly, as a result of seeking answers to the questions: are men and women identical or are they different? What exactly are the differences between women and men, what are their causes and effects? Henceforth, feminist theory emerged.

It was necessary to acknowledge the presence of discrimination against women in society and the conviction that the women's secondary social status is not determined by the sexes' biological differences. All feminists criticize the patriarchy and insist on the need for changing traditional social, political, and personal practices to improve women's social position; they highlight the task of liquidating discrimination in places where biological gender acts as the basis for prejudice (for example, in the labor market, where women's reproductive roles are often a basis for professional discrimination).

Age, interpreted as and an independent dominant, in the opinion of renowned philosopher M. Epstein, becomes "a question of performance, accepted age styling" (Epshtejn 2006). In all probability, studying "age display" will soon become just as popular as "gender display" was not long ago (Zdravomyslova and Temkina 1999; Yarskaya-Smirnova and others 2004). For example, the following definition is possible: Age is a system of interpersonal cooperation, through which a representation is created, approved, confirmed, and reproduced about younger, older/young, and elderly as basic social order categories. Naturally, it appears that age or age self-presentation depend on social and historical context.

Age in this definition also ceases to be ascriptive, while age relations can be considered in terms of socially organized relationships of power and inequality. In a marketing approach framework, inequality in industrial relations can be interpreted as an advantage of experienced and qualified "older" workers over younger; "mentors" and "students."

In modern industry, when the working capital of experience is depreciated and retreats in the face of constantly updated knowledge, the youth occupy a leading place, while the elderly "fall behind" and "decline," especially when talking about ICT and various gadgets; the symbols of information advancement, which we will discuss later on in detail.

Within Talcott Parson's approach, imparting great significance to norms and role behavior, we also see a shift of role repertoire towards expanding the temporary space of "youthful" roles of student, newlywed, and young parent, etc., which was discussed in Sect. 4.1. Youthful roles offer certain benefits associated with less social responsibility, freedom from commitments, ability to manage time and use it for leisure, which in modern society is considered to be individualized possession of important capital or social resource.

The presence of such individual resources gives wide opportunities for inclusion in "consumer society," in which time is valued not as a chance to work more, but as a chance to consume more diverse amusements. Instead of following the cultural-role standard, which recently attributed narrowing and decreasing demand (desocialization) to older people, the new generation of elderly portray themselves as active consumers "who are worth it." The first publications have already been distributed about the "egoistical generation" (Gilleard and Higgs 2014) – a generation that, moreover, is not the least bit ashamed. However, they are not ashamed in particular because they do not want want to yield to younger people's style of consumption, which modern society prioritizes.

In modern Russian society also, based on bad Western examples, a cult of consumption and satisfaction has quickly developed, exalting youth and beauty. This confirms Johan Huizinga's opinion that modern society and its behavior conforms to childishness: game-like behavior models, strength demonstrations (parades, marches), and a cult of physicality (Huizinga 1971). Reassessing values leads to advancing the young as the best of all ages, as a measure of man's "humanity."

This "generational breakdown" (analogous to gender trouble, according to Garfinkel), incites assertion that the elderly population burden is too heavy for the next, less populous generation. This flips the switch from scholarly interest to and actual political agenda, stating simply: "there are too many elderly people." The main measure to lessen the burden, which has already been employed in the USA and Western Europe, is the political decision that pension age, and old age accordingly, now begin at 67 years and in the future might begin at 70.

Therefore, it is not anthropology, physiology, medicine, and other sciences, but government in direct and blatant form acting as an "age maker." Incorporating relationships to power into age relationships dictates that the Western elderly should age more slowly and limit leisure activity, trying to revive the "Weberian approach" towards labor value.

It seems that government officials are considered experts in this case! The contradictory nature of afore-mentioned trends and assessments require that an "age policy" or "aging policy" be developed to replace the traditional "retirement policy" or "senior citizen social service policy." The phrase "age politics" updates associations, just as "Sexual Politics" (quoted from Gapova 2008). Essentially, over the years, feminist sociologists have succeeded in proving that sex and gender are different categories and thereby removed a multitude of women's problems from under the dictation that biological conditionality and showed how much exactly constitutes identity.

Apparently, the specifics of health, educational problems, employment level and labor market position, as well as social roles and status are all connected with socio-economic inequality, more than women's biological features. Admittedly, this question remains in discussion to this day, as it affects the bases of men's socio-economic dominance, although today no one argues with the fact that in the socio-humanitarian sciences (yes, and in natural sciences, as proved by Michael Polanyi (Polanyi 1998) authors are greatly dedicated defending this particular position. Furthermore, gender sociology has long focused on young women's problems – sexuality, pregnancy, childbirth, modern families, and job duties (dual employment) – while the particularities of elderly women's situation is generally unseen.

In particular, this problem relates to the "sandwich generation" of women today who are not just grandparents to grandchildren and parents to children, but are also frequently the children of their own "fourth age" parents. In addition, aged parents require no less care than grandchildren, and acceptable institutional support for a "care-giving" family or an elderly daughter is currently still in the making. The new discourse about exploiting "third age" women has yet attracted little attention from Russian sociologists, with the exception of Elena Zdravomyslova (Zdravomyslova 1999).

In studying aging and old age, we observe phenomena, involving the mutual dependency of biological, psychological, and social factors. Aging is a genetically programed process formed through evolution, in which complex age changes affect an individual on all levels. It is simultaneously a process of destruction and development, involving complex interactive phenomena of decline and advantages and readiness for fight or flight to preserve achieved social status.

Elderly social status in modern society is ambiguous and the question of socio-economic and political stratification and differentiation of age-diverse groups appears quite mixed-up, especially in a country like Russia under conditions of a fragmented and unstable social structure.

In modern aging policy, it is already impossible to reject questions of dominance and submission in intergenerational or age relations. The stable "tendency to treat the old age as starting at the officially established male retirement age continues to prevail in the social consciousness. In other words, society is in no hurry to part with the traditional notions of 60-year-old "geezers" (Shmerlina 2013: p. 183). This just speaks to the desire to preserve "young" population's dominant status in a situation there the number of elderly people is growing.

Modern senior citizens represent a large and diverse community of people, born approximately in the period from 1916 through 1954, i.e. people over the age of 60, seemingly united only by the number of calendar years lived. Age as such can only give us a basis to formally place an individual in a certain age group, but more than that cannot help in characterizing an actual person.

Age studies, conducted within the framework of age sociology, focus on the age structure of society, separate social groups, and the structure's development patterns, as well as the social aspects of an individual's and social groups' age characteristics, transforming in studies of generations and intergenerational dynamics. Indisputably, it is impossible to assign exact features to a specific individual based on age.

However, age points not only the number of years lived, but also to the historical time: the era and social environment in which a person was born and lived. Phylogenesis (a socio-cultural level) is inseparably tied to an elderly person's individual psychological content (ontogeny). In addition, the shadow left by past years and life's peripeteia can say a lot about a person.

Historical and life events form identity, its orientation, appetence, and values. Thus, age and its associated social roles serve as a basis for uniting age cohorts and provoking antagonism from other age groups.

The period of aging must be considered a permanent component of society's social processes. Economic and technical factors, urbanization, and globalization processes substantially determine the content of aging. Such generalization underlines the existence of many variants or the multiplicity of aging, which alludes to the possibility for comparison not just in historical and cultural contexts, but also within the borders of the socium, in which elderly people spend everyday life. Senior citizens' experiences are inseparable from existing institutional aging characteristics: personal development and the development of an aging society are interconnected, the relationship to a specific older person is in many ways determined by social context.

Thus, analogous to gender, the understanding of age is being reinterpreted, as chronologically it can only be objectively determined by entry in a passport or other document requiring birth date. To what degree is this entry essential to personal identity? Usually in modern sociology age is not even mentioned in definitions of "identity" and the accepted model of identity is oriented on the concept of a rational, comprehensive subject, possessing a set of stable values despite the dynamic nature of modern social practices.

Researchers affirm that "although through efforts of deconstructivism, feminism, and other types of postmodernist theories "identity" has lost its inherent essential status as the comprehensive, natural, and unchanging core of personality and has turned into a construct, a never-ending process of reformulating and conversing signs of belonging to a certain "imagined community" (as Benedict Anderson determined), nevertheless, it remains an important concept in interdisciplinary studies (Anderson 1991).

In postmodernist theories, personality is unstable, contradictory, and constantly in the process of formation, while identity is something fragmented and constantly changing. However, in feminist sociology, "citing the multidimensional structure of the identity of modern personality and the political relevance of these identities indicates that the minimum set of differences that should be recognized are class, religion and gender" (Phillips 1992), but for some reason age is forgotten, though it is very important both politically and biographically.

Age also undergoes "postmodernist" fragmentation and reinterpretation within new approaches. We observe not only attempts to partition three, five, or more ages, but also the more radical suggestion by philosopher Mikhail Epstein, who considered that the period of "childhood" has long been differentiated into childhood, adolescence, and youth.

Instead of disputing the differences between ages and contrasts between youth and old age, the philosopher put forth the idea of fractality, i.e. self-similarity of five main ages within each period: "These are metaphors that apply to the ages of human life by their own measure, i.e. base themselves around internal age periodization of the same division of ages that is usually applied to human life as a whole. But where is it possible to get the best division as it has historically developed in very different cultures and been confirmed on both an intuitive and theoretical level? (Epshtejn 2006).

Henceforth, life's course is described as a process of undulating flow, the rises, falls, and repetition of each of the five ages (childhood, adolescence, maturity, advanced age, and old age) in each period of life, for example, "childhood maturity" or "youthful old age."

In modern society it seems that age structuring has noticeably changed; the certainty and consistency of age ranges are not as obvious as in the modern era under the predominance of "uniform time structuring." By way of maintaining linearity, the nonlinearity of life's trajectory is also noticeable. It is acknowledged that traditional structuring and sequencing are still comparatively external in the space of life's separate social groups and people. For someone, this external time structuring is undoubtedly the most important guide for forming a personal biography, but for someone else this orientation may not be as important.

The period before reaching maturity is especially complicated, which leads to pushing away its borders: "Prolonging the age of youth, especially in educated population layers (a writer or artist, as well as a doctor of science is considered young until age 40), generates an "offset median paradox." Individuals are involved later in paid employment or child care responsibilities, but also possess a longer retirement period of life, which in conclusion mixes or pushes away the customary middle life suggested by the employment system in market economics" (Rozhdestvenskaya 2013).

Employment has ceased to be the main identifier of maturity as it has started to vary and has lost many of its "frameworks" ... However, the "adult" group has not yet become "interesting" to sociologists, as it remains "able-bodied," "employed," or "unemployed," but these are more socio-economic than sociological characteristics.

Today, the age denoting the beginning of growing old is determined everywhere and fixed legislatively, since retirement systems exist in essentially all of the world's countries. Thus, the state decides when people become elderly/old/unable to work, based on the condition of the pension system's own finances and resources. However, the population has different priorities.

If modern culture favorably accepts prolonging youth in the form of extended education or postponed marriage, in contrast it meets the idea of pushing back the borders of old age as defined by retirement with clear displeasure. This is expressed by rejection, nearly to the point of mass demonstrations, by populations of various countries that have raised the retirement age, particularly in terms of increasing required insured experience from 35 to 40 years to receive a full pension.

It would seem there is nothing joyful in being "old" (after all, pensions are awarded by old age). No one argues with the fact that changing the retirement age is the most important political decision in terms of age in general, the relationship of life's age periods, and intergenerational cooperation, etc. However, this increase is discussed only as the manifestation of authority's repressiveness and developing modes for socio-economic suppression of hired workers, i.e. the majority of the population. People forget or do not consider that "comparative deprivation associated with retirement frustrates the symbolic order" (Sztompka 1993: 378), exhibiting repressiveness in interactions on the micro-level.

But this is a one-sided view from modernism, from a society where "to work is to pray," the opinion about the sacred nature of work prevails. Yet, in contemporary "consumer society" there are other values. Life itself, i.e. lifetime, has become a subject of consumption, while work feels like an obstacle in the way of this consumption. Providing a means to consume things, employment takes away the more valuable: time, the time of romance, which never gets old, romance with oneself....

Opinions of members of the social sciences about the social situation of old age also diverge. Many see the main problem of old age in the abandonment, desertion, and unnecessariness of older people. Others see the potential for "decisive self-realization completeness" in old age, or the period in which a person discovers either the great meaning or meaninglessness of life lived.

In Russia, "passport age" or "visually identifying age" remains a major factor for the government, society, inner social circle, and elderly people themselves, sharply limiting professional itineraries and life plans. Thus, the functionalist tradition comes to fruition, where individuals were interested insofar as they fulfilled normal requirements of the social structures of "disciplining society" (Mihel 2003; Foucault 1999).

Within framework formed by experts, first and foremost medics and age psychologists, as well as the social opinion on "readiness to resign," the life horizon sharply narrows, while life itself "in older age" is considered from the point of view of living out one's days. Although institutional frameworks in any society present a corridor of opportunity for individuals, in modern society this corridor is more of a space for independent choice and identity multiplicity.

4.2 Age and Time, Social and Individual, Generations and Conflict

Industrialization and urbanization are widely believed to have rapidly reformatted time for society. Uniform time-measurement, subjected to socio-economic life, developed over the past 3–4 centuries. In the modernist type of society, the majority of the population had an involuntary life schedule, which can be interpreted as both organizing and oppressive. The length of the time period for receiving education and employment, the designated time for marriage, and other life events very strongly differed amongst different socio-economic groups.

However, many processes that are becoming universal, for example, a prolonged educational period, unified the age structuring of life and created prerequisites for life's path to unfold linearly. Receiving an education was considered a prerequisite to professional qualification and inclusion in the labor market, at the threshold of which there was no cause to linger. Marriage was just as voluntary-compulsory in nature, also separating youths from adults.

The theory of age stratification, associated with George Herbert Mead's concept of symbolic interactionism, corresponds to traditional views. He describes society as an aggregate of age groups that differ in abilities, roles, functions, rights, and privileges that are conditioned by age. Chronological age is the basis for stratifying and explaining different generations' characteristic features.

American sociologists N. Howe and W. Strauss offer a close look associated with time of birth. From their point of view, members of six generations currently live in modern Russia: The Greatest Generation – the oldest people, born from 1900–1923; The Silent Generation – younger retirees, born from 1923–1943; the Baby Boom Generation – the generation of people born in 1943–1963; Generation X – the middle-aged generation (born 1963–1984); the Millenial Generation – the young generation just now beginning life's path as adults (1984–2000); the Homeland Generation – the generation of people born in or after 2000 (Howe and Strauss 2000).

The prominent sociologist V. Semenova has made a large contribution to generational analysis in Russia (Semenova 2014). She segments 4 generations in modern Russia: the "wartime" generation, whose core members possess 60–70 years, the "pre-perestroika" generation, 45–55 years (further, 50-year-olds); the "reform" generation, people ages 30–40 years (further – 30-year-olds); and the "post-reform" generation, people ages 18–25. The author notes a turn in sociology towards text analysis and developing interests in the evolutionary processes, including biographies (as Sztompka calls it, the "sociology of becoming"), i.e. interest in studying culturally-historical aspects of social processes and occurrences (Semenova 2014).

Modern social studies, as mentioned above, along with discussions about age differences and the contrasts between young and old, have pushed forward the idea of fractality, i.e. the self-similarity of five main ages within each age (Epshtejn 2006). A multitude of identifying events, like receiving an education, marriage, employment change, and so forth are sprinkled and/or repeat throughout life's course.

In everyday life, people have become long accustomed to newlyweds at the age of 50–60 or "eternal students." The popular idea of "life-long learning" not only affirms the importance of relearning/learning as much as possible, but also establishes the normalcy of an eternal student, i.e. a never fully-maturing "pupil," on a social status level. Henceforth emerge the fractal phases of "old childhood" and "youthful maturity," etc.

"Reshuffling ages is especially characteristic for modern societies where social responsibility all the more often shifts onto the young; schoolkids in older classes already experience sometimes unbearable pressure (which leads to increased adolescent suicides). In contrast, this yoke is removed from the shoulders of 50-60-year-olds: social inertia or acceleration achieved in midlife lightly pushes them comparatively further from level to level, allowing them to newly, or even for the first time, relive youth in the transition from adulthood to old age" (Epshtejn 2006).

We think that many people will not agree with this idea, as they have for some reason "insufficiently accelerated" by old age, meaning in more positivist language, they didn't acquire adequate education, qualification, adequate social capital, and have not saved money for a peaceful old age or peak self-realization....

How is it possible to determine the level of a person's dependence or, in contrast, independence from the external age noted in one's passport and from cultural stereotypes? Disagreements of chronological, biological, and social ages have long been recognized in the field of modern medicine, which has learned quite exactly to determine a person's real age and give corresponding recommendations for prolonging life and improving health.

Employment services, according to press messages, are now reoriented on seeking appropriate work for senior citizens (Topilin 2014). It turns out that life-determining events in youth and old age are essentially the same. On an individual level, the question remains of how to recognize the level of a person's dependence or independence from one's externally-fixed passport age and cultural stereotypes.

At the beginning of the twenty-first century, class and even gender conflicts seem to be already less functional descriptions of actuality in theoretical sociology and are replaced by a focus on generation conflict. As noted above, Russian sociologists also note the opportunity for valuable disagreements in relation to intergenerational conflicts. We agree with Valentina N. Yarskaya's opinion, that "interest in working with the generation problem in mass consciousness and the scientific community stimulates orientation on radical and rapid changes through striving for fairness in resource distribution, and through global changes in modern society's age composition" (Yarskaya and Bozhok 2014).

Even with an untrained eye, generational conflict and aging rejection can be seen everywhere in daily life. Older women particularly are particularly affected in this situation. Here, the solid patriarchal traditions of our society prevail over tolerance, and elderly women are especially noticeably displaced into the sphere of deeply personal life, where they in turn are fit only to raise grandchildren and sort out the best variety of vegetable oil, as in the well-known TV ad about which oil a good housewife uses. Even former Hollywood stars age, in spite of the dream of eternal youth and beauty that they embody.

The Social Sciences are filled with varying opinions, from complete acceptance to complete rejection of the possibility of generational succession in a rapidly-changing society. Some authors give basis to the idea of a global, universal, and constantly-widening divide between generations. Others insist that perceptions of increased intergenerational differences are illusory, and that intergenerational cooperation has a pendular nature in which conflict periods alternate with succession periods. The general position of the afore-mentioned standpoints is to adopt tolerance and continuity in relationships as a fundamental condition of societal development and the need for careful evaluation of the rates of socio-cultural renovation.

The "new social question" consists of updating the intergenerational contract, which supports the elderly and invests in the young for stable and fair conditions (Albertini et al. 2007: 319). This movement away from class to generation became possible thanks to the success of the modern Welfare State, which created opportunities for increasing age-diverse requirements and obligations and turning to the elderly as a major client-base for this type of state. Existing demographical changes, i.e. low birthrate and increased lifespan, create prerequisites for new approaches to redistribution.

The further a "social contract" should not be built on regulating relations of working (employed) and no-longer working (retired) people, the more it should become a more complex construction, considering that the mass arrival of women in the workforce has made employment and difficulties in entering and exiting the labor market more flexible.

However, young people relate to the necessity to pay taxes to "support retirees" fairly skeptically, finding fewer and fewer rational reasons to respect this group. Even working retirees find themselves "under pressure of rejection," as justly noted by A. Smolkin, the work of watchmen, conductors, and attendants are connected by the constant need to "check," invoke responsibility, and thwart violations contributes to eroding respect toward older persons and disseminating practices of avoidance in relationships with them (Smolkin 2010). Thus, a field of potential interactional conflicts is formed.

Older people, gaining new social identities in the modern, rapidly-changing society, accept "theirs" from the point of view of loss and gain over years of change. For some, this perspective formulates as: "we are those who have lost due to reform;" for others, it is: "we are those who make our own fate." Age forms a fairly wide radius around social consolidation ("rallying confidence," as put by R. Inglehart and P. Ricceur), but also possesses the powerful potential of social confrontations and conflicts, as displayed in 2005 after Russian Federal Law FZ No. 122, "On Monetizing Benefits," came into effect. But authorities learned their lesson and now act very carefully in terms of enacting changes in the pension system's infrastructure, limiting them to "minor repairs" and other instances able to evoke social-age consolidation.

Thus, modern Russian authority was formed in the late 1990s on the wave of the population striving for order, while today a significant portion of the population calls instead for "security." The state, according to the research of G. Tulchinsky in 2012–2014, answers this request. The transmitted image from the state's side is aimed at the value axis -- "security," protecting the population from threats – and the axis of informal institutions of "culture" supported by tradition and habit. In addition, internal Russian policy frequently changes depending on the value axes of "culture" and "rights," reacting to changes in society's climate (Tulchinskij 2015).

In our view, it is obvious that the state, offering security and tradition as a priority, is oriented on the values of older age groups.

Intergenerational conflicts have always been depicted in artistic and philosophical literature. But in modern society, youth's symbolic domination in the space of mass media contrasts with the traditional economic and authoritative balance of the older group, with noticeable value and technological disagreements, as forecasted by M. Mead (Mead 1970).

The elder's domination when led to absurdity is the theme of renowned author Vladimir Sorokin's story, "Nastya." Instead of Baba Yaga and children from Russian folklore, his story features a band of cannibals devouring the girl Nastya on her 16th birthday. In addition, he does this in a focused and systematic way ("all daughters in area were already eaten," remarks one of the characters), when "machine-usage is far greater than insanity, in that subsists the novelty of experiment" (Pyatigorskij 2001).

Yulia Kisina's story, "The Lithuanian Hand," seems no less absurd in relation to the subject of death and elderly domination, just with a different mark. The story is set a flea market and women sell mummies of their dead, WWII-veteran husbands in installments. The mummified characters are endowed with magical powers: to heal the sick, cause damage/decay, and help with studies. They are loved, protected, and even doctored (if they suddenly start to get moldy). Whole and dissected into hands and heads, the mummies are served by the person that feeds them, leaving the family without a breadwinner (Kisina 2005).

Real life conflicts are sometimes not inferior to those in absurd literature. Sitting for half a day at the Pension Fund reception, an observer notes discussion of intergenerational conflicts due to the diversity of retirement plans, over acknowledging merit in the form of discounts and compensation, and, of course, due to pension size. One of this book's authors observed that one discussion even escalated into a fistfight between two very elderly men.

Thus, on the one hand, contemporary elderly people are geezers in dying-out villages, without a remote shadow of ICT or even a television, since they lack electricity. On the other hand, there are separate social groups that are well-provided for in old age, among of which are former state servants and deputies, who have any medicinal, social, or cultural services at their disposal, regardless of cost.

But society cannot exist as it is without substantial interaction between people of different generations and the desire to coordinate interests. In turn, intergenerational relations marked by intolerance and inflexibility, concluding in "absolute updating" or "absolute divide," cannot be positive, as they risk the collapse of a united social and historical space/time.

4.3 Life Plans

What unites our existence into one, seemingly indivisible whole? In this whole, are there only quantitative growth differences or is a person incommensurable in different ages; is one body a bearer of transformations, in which three or five different people live? How much does consciousness depend on our "fleshy casings?"

The social determinants of the age phases structuring a person's life path are connected to the influence of highly differentiating multi-leveled social processes. Accordingly, German sociologists H. P. Blossfeld and J. Huinink note the following influences:

- the life paths of other people, with whom a person maintains a close personal relationship: parents, partners, children, and friends, etc.;
- social institutions and organizations (intermediate sources, state institutions, labor organizations);
- the living conditions in social and regional contexts, where an individual spends life or which fluctuate over time;
- existing and changing social structures and historical events that represent socio-cultural, political, legal, cultural, and economic organization conditions, etc. (Blossfeld and Huinink 2001).

However, all of these processes and influences require a certain amount of reflexivity from the individual and responsibility for oneself. To what degree does the population think about planning its own aging, and about its life scenario and responsibility for it, in general? It seems that personal participation in life planning is still minimal, while there are contradictory relationships with the state, as people berate it and await a resolution for their own problems. Russian senior citizens have a very high sense of externality, primarily in matters of health preservation. Elderly people associate quality of life with health in particular.

Older people themselves interpret continuing paid employment after retirement age as simply compulsory due to low pensions. Even the low retirement age rarely valued as a Russian advantage, because "everyone is indebted to the elderly, but do very little for them." However, it is necessary to evaluate increasing the retirement age from more than just an economic perspective. Raising the retirement age would create motivation to live longer, because low work experience requirements and low retirement age create prerequisites for exhibiting such a "short life outlook."

J. Huizinga noted long ago that modern society's age priorities are shifted in favor of youth, while the younger generation lives in short, temporary distances and does not set long-term goals, instead simply transitioning from one life "project" to another. Is it possible that old age, the time when people desire to lengthen and prolong plans and relationships, is now starting to prevail?

Unfortunately, the mass consciousness in Russia, ridiculing the gerontocracy of late socialism, faces extreme displays of "young power," and now aims to go "forward to the past," but these wave-like processes do not sustain the idea of the meaninglessness of different versions of ageism.

Chaotic transformations in recent decades also do not promote forming long life plans and reflexive biographies (Adam et al. 2000). It is becoming natural for the modern person to think in terms of risks and security from any type of risk. However, the main risk seems to be uncertainty, primarily, the lack of socio-economic strategies in modern Russia that are both comprehensive and comprehensible to the general population.

This exacerbates the impression that the older generation's increasing size is a dangerous risk to society. Although, by a large number of parameters aging only shifts accents from current areas of social regulation to new ones. However, prolonged aging requires that each person be included in developing his or her own life scenario, the risks of which are individualized, as security is difficult to plan "from above," in terms of the government's abilities.

In the early 1990s, older people themselves unexpectedly and almost simultaneously discerned the unsuitability of social interaction norms that were acquired in childhood, losing their orientation in social space. The older generation was formed in conditions where individual activeness was to be aimed at incorporating into social structures and adapting to externally-given environmental properties. But, insufficient attention is paid in the discourse of social science, daily life, and the media to opportunities for overcoming this situation through dis-adaptation and incorporating new life models.

E. Kosilova believes that the older generation compromised itself politically and morally in our country as well as in many others. In other countries, where there isn't such sharp conflict, young people still sense that their problems with the socium differ from those of older people. Again, in the philosophical sense, problems may remain the same, but society examines these problems through discourse. Discourses themselves have changed in contemporary generations (Kosilova 2014).

Interactive skills play the most important role in a person's social adaptation/re-adaptation, since social problems in a given context appear to be indicators of adaptation violation. However, there is a trend in sociology, more so in social work, to consider the social environment as something external and, in the Russian context, hostile to a person. But from a systemic viewpoint, the world is not something foreign, lying outside of us, but is instead the summation of our temporal perspectives and opportunities.

Thus, demand for social work grew rapidly against the backdrop of accentuating the violation of ideas, values, and economics, etc., which was easily built into the "pitying humanism" that Russians find accustomed. The famous sociologist N. Tikhonova noted long ago that state policy is not aimed at supporting the low-income but working layers of the population. It is focused on people who are unemployed and devoid of childcare responsibilities; in a word, state policy is concentrated on supporting the "social sea bottom" (Tikhonova 2002).

The development of social work in Russia, and even more prominently in Western Europe, provides a multitude of examples of how ineffective social entities can win in the competition for survival, because they spend a minimum of personal resources and successfully attract institutional resources. Literature on the subject already points out the rationality of passive adaptation for people living in the socially-depressing environment of small settlements. These people do not want to change their situation because, while there is no work available, this is where their house and social connections are – "everything familiar."

"Breaking points" arise when entering and exiting a full and independent social life. Yet a mandatory and regulated exit, for example, retirement, emerged recently historically to benefit growing masses of industrial workers. Is such an exit necessary, i.e. exclusion from work relationships, similar to the way that inclusion in these relationships was mandatory in youth? Apparently not, in a society of knowledge and qualified labor.

But long life plans are characteristic of more educated parts of society, along with long periods of education followed by employment. We emphasize that education here is not equivalent to training, but is much wider than strictly preparing for a certain type of work.

A situation arises among the less-educated where "time is bigger than life" or than interesting lessons that fill life, which evokes total displeasure... Sociologists' opinions also diverge. Some see the main problem of aging in abandonment, neglect, and unneededness. Others see the onset of old age as an opportunity for "fully-decisive self-realization".

It follows to mention that an individual's social activeness is closely tied to social well-being. Our studies of elderly people, beginning in 2003 in the Moscow administrative region of Saint Petersburg (Otchet 2004) showed that senior citizens frequently forget that a direct dependence exists between social activeness level and actual engagement and satisfaction with life. The sense of opportunity influences the path of events and changes in personal life is interconnected with positive acceptance of change. Unfortunately, research shows the opposite situation: social passivity amongst the elderly, mistrust, apathy, and, consequentially, a negative emotional context and general dissatisfaction with one's surroundings (Kiseleva 2010).

Social passivity is also associated with the external direction older person's personality takes. This means meaning reconsidering the result of one's experience not as the conclusion of purposeful efforts, but rather as an attribution by force of circumstance. The tendency to ascribe responsibility to external factors (surrounding environment, fate, or chance) is characteristic of older people and, naturally, does not bolster motivation to overcome difficulties. The elderly population displays similar externality in paternalistic installations, orienting on state interference in resolving any problem of old age.

Throughout this project, we also studied older persons' identification resources. We employed T.Z. Koslova's classification (Kozlova 1994) when processing answers to the question, "Who am I?" (M. Kuhn and T. McPartland's self-identification test).

In our sample, the study demonstrated a full lack of self-identification in terms of political views, personal abilities, education level and material well-being, external data, religious and ethnic affiliation, as well as identifying with one's homeland. The only instances of identification were in relation to family position, people of one's town/country, with humanity, with people based on living place, and on interests. Main identification resources are kinfolk (mother, grandmother), friends, retirement status, health condition (handicapped), personality traits (kind, honest, practical, level-headed), mental features (optimist, idealist), and lifestyle (active, resting in resorts, love traveling).

These are the leading life spheres in which an older person is most successful and motivated. Character qualities lead amongst identifying preferences. We only encountered two examples of negative self-identification: being dissatisfied with oneself, disliking oneself. Essentially, seniors have a high positivity level associated with personal identity. Here, we are dealing with a person's "displayed image" – "I" represents that which a person chooses and wishes to show.

The fact that amongst identifying groups, daily interactional groups (family, friends, people of the same generation, work friends, and colleagues) dominate over imaginary and fabricated communities (humanity, Russians, and Christians), attests to people's retreat into private life, within the framework of immediate cares where an elderly person feels "at home" and protected. This is a display of the so-called "snail effect," or incorporealization, where an older person concentrates interests, attention, and care on problems within his or her close social circle (Krasnova 2000).

The circle of social self-definition and self-realization is narrowed down to primary groups – to environments that subjectively deemed most acceptable. Thus, the leading adaptation mechanism for older people is traditional – through family and close friends.

We also witnessed difficulty in subjects' social thinking and reflection, their "closed consciousness," and localization of life space as in the literal as well as figurative sense. Here, in all likelihood, the only resolution for the circle of psychosocial problems older age is to develop territorial solidarity through local government institutions and communitarian work.

On the other hand, the dwindling sense of closeness, cooperation, and solidarity within all groups reflects a psychological condition associated with experience in isolation. Society, in turn, exhibits an inability to offer collective forms of solidarity to older persons.

In the absence of group identification, social structure conceals the reason why we do not yet see increased social activeness amongst the expansive group of elderly people, possessing impressive time resources.

An individual identity demonstrates social activeness not only fulfilling personal interests, but more so in fulfilling the interests of a specific social community, in which a person is objectively integrated and with which he or she subjectively identifies. In modern Russia, visibly, there are few inclusion mechanisms and suitable communities for the senior citizen, much less faith in such structures.

The older person lives locked in a circle of everyday life, where he or she clearly submits to the absence of bright feelings and impressions and to the desire to shield oneself from the outside world's problems. The lack of a minimal stress factor associated, for example, with the opportunity to master new types of activity or diverse forms of activeness, encourages the "snail effect," narrowing life's trajectory plane. Noticeably, this is one of the reasons we exposed in studies during 2014–2015 of senior citizen's very narrow "leisure" interest profile in mastering ICT (see Chap. 5 for more detail about continuing education and ICT training).

Dearth of objective reasons for anxiety – so-called "healthy" anxiety – leads to stagnation in the emotional sphere and decreased spiritual sphere quality, which is inevitably reflected in somatic health. Thus, the effects of losing loved ones in old age frequently translate into the loss of life's meaning in general and the inability to fine a new one.

***

The goal of an aging policy is to become more compassionate toward people's diverse life scenarios. In modern society, elderly people should receive more "rights to authorship," acknowledging their different life strategies, trajectories, and types of self-identification in terms of primary biographical blocks. Meanwhile life plans seemingly must teach or help to build and develop in any age.

Here, we see the specific task of humanizing existing social work and protecting senior's interests from the too-calculating liberalization of these social services, which, perhaps compulsorily, the state requires.

References

Adam, B., U. Beck, and J. Loon. 2000. *The risk society and beyond: Critical issues for social theory*. London: Sage.

Albertini, M., M. Kohli, and C. Vogel. 2007. Intergenerational transfers of time and money in European families: Common patterns – different regimes? *Journal of European Social Policy* 17 (4): 319–334.

Anderson, B. 1991. *Imagined communities: Reflections on the origin and spread of nationalism (revised and extended. Ed.)*. London: Verso.

Bauman, Z. 2004. *Wasted lives. Modernity and its outcasts*. Cambridge: Polity.

Blossfeld, H.-P., and J. Huinink. 2001. Lebensverlaufsforschung als sozialwissenschaftliche forschungsperspektive: themen, konzepte, methoden und problem. *Bios* 2: 15.

Bourdieu, P. 1998. Struktura, gabitus, praktika. *Zhurnal sociologii i social'noj antropologii* 1: 2.

Eisenstein, Hester, and Alice Jardine, eds. 1990. *The Future of difference. Responsibility*. New Brunswick: Rutgers University.

Epshtejn, M. 2006. K filosofii vozrasta. Fraktalnost zhizni i periodicheskaya tablica vozrastov. *Filosofskij kommentarij. Zvezda* 4.

Evans, J. 1995. *Feminist theory today: an introduction to second-wave feminism*. London: Sage.

Foucault, M. 1999. *Surveiller et punir: Naissance de la Prison*. Paris.

Gapova, E. 2008. Kejt Millet: lichnoe kak politicheskoe. *Neprikosnovennyj zapas* 4 (60).

Giddens, A. 1990. *The consequences of modernity*. Cambridge: Cambridge University Press.

Gilleard, P., and P. Higgs. 2014. *Ageing, corporeality and embodiment*. London: Anthem Press.

Habermas, Jürgen. 1984. *Theory of communicative action. Volume one: reason and the rationalization of society*. Trans: Thomas A. McCarthy. Boston, MA.: Beacon Press.

Howe, N., and W. Strauss. 2000. Millennials rising: The next great generation. In *Cartoons*, ed. R.J. Matson. New York: Vintage Books.

Huizinga, J. 1971. *Homo Ludens: A study of the play-element in culture*. Boston: Beacon Press.

Kiseleva, Ye. 2010. *Utraty v pozhilom vozraste I adaptatsiya k nim. Pozhiloy chelovek v obschestve. Pod red. Grigoryevoy. St.* Petersburg: Euopean House.

Kisina, Y. 2005. Litovskaya ruchka (rasskaz). In *Russkij rasskaz XX veka. Antologiya*, ed. Sorokin, vol. l, 551–554. Moscow: Izd-vo Zaharov.

Kosilova, E. 2014. O vzroslenii v multifaktornoi kulture. *Otechestvennye zapiski* 5 http://www.strana-oz.ru/2014/5/o-vzroslenii-v-multifaktornoy-kulture. Accessed 6 Dec 2017.

Kozlova, T. Z. 1994. Samoidentifikaciya nekotoryh socialnyh grupp po testu «Kto ya?». *Social'naya identifikaciya lichnosti. Kn.1*. Moscow.

Krasnova, O.V. 2000. Babushki v semie. *Sociologicheskie issledovaniya* 11: 108–116.

Mead, M. 1970. *Culture and commitment - published for the American Museum of Natural History*. Garden City, N.Y: Natural History Press.

Mihel, D. 2003. Vlast, upravlenie, naselenie: vozmozhnaya arheologiya socialnoj politiki Mishelya Fuko. *Zhurnal issledovanij socialnoj politiki* 1: 91.

Otchet o nauchno-issledovatelskoj rabote po teme. 2004. Izmenenie statusa pozhilyh lyudej, ponyosshih utratu blizkih, ih socialnaya adaptaciya i reabilitaciya. In *V ramkah Federalnoj programmy "Starshee pokolenie"*. Reg. No 01200408312. Moscow: VNTI Center.

Phillips, A. 1992. Democracy and difference: some problem for feminist theory. *Political Quaterly* 63.

Polanyi, M. 1998. *Personal knowledge: Towards a post-critical philosophy*. Abingdon: Routledge.

Pyatigorskij, A. 2001. Igor Smirnov i Vladimir Sorokin. *Novaya russkaya kniga*: 1.

Rozhdestvenskaya, E. Y. 2013. Biografiya kak socialnyj fenomen i obekt sociologicheskogo analiza. Avtoreferat diss. dokt. soc. nauk. Moscow.

Semenova, V. V. 2014. Strategiya kombinacii kachestvennogo i kolichestvennogo podhodov pri izuchenii pokolenij. *INTER* 8.

Shmerlina, I.A. 2013. Liberalizaciya stareniya: teoreticheskie illyuzii i ehmpiricheskie anomalii. V. 1. *Monitoring obshchestvennogo mneniya: ehkonomicheskie i socialnye peremeny* 3 (115): 165–175.

Smolkin, A.A. 2010. Bednost' i social'nyj status pozhilyh lyudej v sovremennoj Rossii. *Monitoring obshchestvennogo mneniya* 3: 186–199.

Sztompka, P. 1993. *The sociology of social change*. Oxford: Wiley-Blackwell.

Tikhonova, N.E. 2002. Socialnaya eksklyuziya v rossijskom obshchestve. *Obshchestvennye nauki i sovremennost* 6: 4–21.

Topilin, M. 2014. Starshee pokolenie sejchas drugoe. *Rossijskaya gazeta* 01.11.2014.

Tulchinskij, G.L. 2015. *Rossijskaya politicheskaya kultura: osobennosti i perspektivy*. St. Petersburg: Aletejya.

Yarskaya V. N., Bozhok N. S. 2014. Perspektiva pokoleniya. Dvizhenie istoricheskoj rekonstrukcii // Izvestiya Saratovskogo universiteta. Novaya seriya. Seriya: Sociologiya. Politologiya. T. 14. No 2. pp. 19–26.

Zdravomyslova, E.A., and A.A. Temkina. 1999. Socialnoe konstruirovanie gendera kak feministskaya teoriya. In *Zhenshchina. Gender*, ed. Z.A. Hotkina, N.L. Pushkareva, and E.I. Trofimova, 46–65. Moscow: *Kultura*.

Chapter 5
Health, Adaptationary Medicine, or Healing Sicknesses?

Aging has long been considered one of the most critical problems of modern society and social policy. Furthermore, the social experience associated with life in an aging society does not yet exist, but thinking in terms of risks and threats is entirely widespread. In particular, people speak about aging as a risk in many international documents over the past 20 years: "The world draws close to a crisis summoned by its aging… The proportion of elderly people in the population grows quickly, increasing the economic burden on the young" – these are the usual statements when speaking about aging.

Or still more frightening: "World aging will disrupt the economic structure, redefine family relationships, force world policy reform, and even change the geopolitical order of the twenty-first century" (Global 2000).

Naturally, this "moral panic" hinders experts as well as government officials developing rational relationships toward the aging population. Seniors' own fears are that sickness, weakness, and helplessness are unavoidable in old age feed into existing stereotypes about the impossibility of healthful aging. "Subjectively constructed weakness" by older people themselves incites development of "care optics" amongst subjects, government, health care, and social services, that would be more acceptable for relationships to people of the 4th age, not the 3rd (Rogozin 2012a, b). In short, different sides of the structuring-structuration process constantly change places. However, this feeds into the fear that until recently developing a country's healthcare system was built around a model designed to help young, able-bodied people in critical or episodic conditions, and doesn't answer the needs of older patients with chronic illnesses.

In 1948, the World Health Organization (WHO) suggested that health be understood as a state of "full, physical, spiritual, and social well-being and not just the absence of illness and physical defects." Today this definition undergoes criticism in connection with its interpretation of health as a statistic category, and also because of the impossibility of developing unified well-being standards for different social groups.

© Springer Nature Switzerland AG 2019
I. Grigoryeva et al., *Elderly Population in Modern Russia*,
https://doi.org/10.1007/978-3-319-96619-9_5

As a result, the dynamic interpretation of health is used more actively in social studies; the representation of health through a process of individual/social group adaptation to life environment. We will now try to understand this in more detail.

5.1 Aging and Health Loss Risks

Today, the process of population aging has acquired a global nature, affecting all countries. Even in developing countries with a high birthrate level, after the year 2025 the elderly population is projected to include more than 20% of citizens, which is connected with rapid population aging in the most thickly-populated world's countries: China and India. The rapid increase in the number of older persons in a population's structure began in developed countries, but is now present in almost all countries, and has resulted in an increased number of publications about the "threat (risk, challenge) of aging." This situation can be compared to the period in the mid-1960s–early 1970s when sociologists actively studied youths after the "1968 revolution".

Demographic changes were perceived as crises, although in a significant number of parameters they only shift accents from existing areas of social regulation to new ones. In particular, they constantly emphasize the risk of possible increased government and society expense in the field of healthcare and social policy, or accumulating elderly poverty. We recall that the concept of "social risk" is historically associated with the risk of income lost for hired workers losing the opportunity to participate in the economic process.

Social risk reasons and factors frequently have an objective nature in relation to a single person: the possibility of unemployment, loss of ability due to an accident, sickness, and the most unavoidable reason – aging, i.e. the loss of ability as an effect of aging, etc. In this type of case, neither bachelor worker, nor family man, nor a person included in a team of workers is able to prevent adverse events and reduced quality of life, i.e. poverty. But these risks in particular are massive and traditional. A branched system of social insurance was created to prevent them, laid forth in the 1980s by Otto von Bismarck, the Chancellor of Germany.

Contemporary authors, first and foremost Ulrich Beck (Beck 2000) and A. Giddens, have suggested the popular concept of a "risk society" that examines social risks as the result of complicating social systems and increasing the number of unintended consequences from their interactions. Risks originating from hyper-complex technological and socio-economic systems escape the control of not only individuals, but of any organization, including even the government.

Russian authors have published a number of works on this topic (Zhuraleva 2006; Yanickij 2003). Thus, in the most generalized view, social risk is understood as the potential commencement of all significant negative events or actions together that can harm an individual, social groups, territories, societies, and the government as a whole.

Russian economist V. Roik also notes that the rising social risk level is unreasonably considered the unfortunate coincidence of circumstances or ignorant reform; it is a legitimate, natural process in our conditions, "the flip side" of increased economic freedom. From thence comes the undeniable necessity for creating mechanisms and forms of social protection that will be adequate for new conditions in society and for newly coordinating workers' interests, enterprises, state, and society (Roik 2011).

In modern Russia, there is an entire system of state standards for classifying various anthropogenic risks. A system of regulating social risks was also created, which encompasses the Emergency Control Ministry (EMERCOM). But the system of mandatory social insurance i.e. insurance from social risks, is far more important and has existed since Soviet times, although it is now being reformed due to new socio-economic conditions.

To sum up, any social action is potentially risky. Inaction or refusing to perform an action can pose no smaller risk. If in 1990 the Russian philosopher A. Algin argued and questioned: "The purpose of risk problems from a philosophical point of view can be formed in the Kantian spirit: Is it possible that the risk-assessment vision of society, human history, and action are processes of regulating different areas of social life?" (Algin 1990: 37), then we have the right today to answer that risk assessment is the only type of vision.

Risk evokes the effects of distrust towards the social system, since decisions are based on trust, i.e. the predictability of reciprocal action and the minimal number of unpredictable or accidental outcomes (Merton 1936). Thus, Giddens treats trust as a necessary condition for decreasing or minimalizing risk, because in a situation of trust, action alternatives must at least exist. Not only is trust in others important, i.e. assuredness, but so is trust in oneself and the ability to take ownership for one's own behavior. Consequently, the majority of "accidents" penetrating people's activity are created by people themselves, not given by God or nature.

From this updated understanding, risks are being reconsidered today, for example, the risk of health loss. In the four-factor health model that has been generally accepted in recent decades, "individual lifestyle" carries more and more weight in comparison to those connected with medical intervention and treatment. As a result, medicine in general should be preventative, not curative, and each person should first and foremost look after his or her own health. Instead of the "Medical Nemesis," punishing people with increasingly sophisticated and early diagnostics, this understanding of the interaction between health and healing suggests a "healthy lifestyle" and supports the organism's adaptive capabilities.

Accordingly, WHO's "negative" definition of health as "the absence of sickness" can (and even should!) be replaced by the notion of health as an individual, social group, and society's (when talking about communal health) process of adapting to changes in the surrounding natural and social environment. In this case, the possibility of disrupting adaptation to the environment becomes a health risks, because it can indicate low endurance potential, as well as super-fast social transformations. The significant deficiency of individual resources is a risk in the circumstances of rising social demand for socially-passive individuals, which extends to a great number of elderly people. Reasonably, hygienists and social workers should provide key support in terms of this understanding of health risks, not doctors.

We also note that risk replaces the inescapable antiquated New Time's Fate and Fear of God as a key life category. Both of these things were outside of human control, in general. In relation to the age of industrialization's effects, developing social insurance in essentially all countries generated confidence in the fact that risks can be managed and that a rational insurance system can conquer causes of income and poverty loss, as well as sickness, aging, and occupational injuries, etc.

However, today we must talk about extensive risk environments, for example, poor regions even in wealthy countries. Naturally, there are many such regions in a country as enormous as Russia, where interregional inequality has only increased in recent years (Zubarevich 2015). As such, "the concept of risk becomes central to society, which breaks with the past, with traditional means of action, and greets the problem-filled future head-on."

The new situation truly requires that new approaches be developed for planning population health and for the population's participation in this process. Meanwhile, it is important to "redeploy" resources for the older population's use. In Russia, a decent pediatric medicine infrastructure remains from Soviet time – children's clinics and specialized health Universities, which are turning out pediatricians of diverse specializations. Clearly, the first step is to partially reorient this infrastructure (an important measure to calculate!) to open more geriatric offices or create specialized clinics to prepare more students as geriatricians.

For now, there is a tangible deficiency of qualified departments for working with elderly people, or even the full absence of specialized aid. The situation is not progressing because there is not enough knowledge from neither "above," nor on the most massive level. This leads to a tapering view of elderly people's problems, as E. Pushkova, the first director of the Urban Geriatric Center in Saint Petersburg, cautioned in the early 2000s. Fear-mongering stories prevail over the social consciousness, more than sober analysis of the adaptation measures society needs to change its age structure.

It is important to note that older people's somatic and mental health often become a problem not just for society's perception, but for their own self-perception. Mental health is the state in which a person can realize his or her own full potential, deal with the usual life stresses, labor productively and fruitfully, continue to enjoy life, and contribute to one's community (Mental 2014).

Modern society, hypostasizing constant update like the majority of people, has been unprepared for old age as a particular life stage, the experience of which is determined primarily by the content that will fill daily life. Naturally, educational programs for the elderly should serve the purpose of such preparation. The oldest educational programs for people of the "third age" exist in France. These programs were oriented on providing knowledge about the particularities of elderly health and older persons' behavior in situations of sickness.

Programs were developed in relation to several directions: the account of patients' statements about health; engaging a patient to cooperate with a doctor; the aim to make healing and educational intervention more individualized; and the account of physical, psychological, and personal particularities. The main goal of this education is to change the behavior of senior citizens in relation to their health (balanced nutrition, physical and intellectual activeness, and so forth), i.e. to do everything possible to promote healthy longevity, which few people in Russia currently believe is possible.

Even scientific and scholarly publications (Yatzemirskaya and Belenkaya 1999) emphasize the biological aspects of aging associated with physical and mental decline, sicknesses, and end of life, which is the effect of a completely antiquated view on health, illness, and human life. At one time, health and life were perceived as gifts from God. Changing or improving health was considered sinful from that viewpoint, as it interferes with divine providence and sickness is one of the components of original sin, which dwells in every mortal. Sicknesses or ailing people were accepted in Christianity and, moreover, believed to be the norm: "Flesh is weak, but spirit is strong..." This view leaves its mark in discussions about the fact a person must inevitably lose health with age.

Meanwhile, it is obvious that in the middle of the nineteenth century (later in Russia), average life expectancy began to grow, sickness retreated before scientific progress, and death became demystified. The trend of romanticizing illness, for example, consumption (tuberculosis), which ended the lives many literary heroes and even more heroines, fell to the past. In the first half of the twentieth century, the totalitarian-military regimes of Germany and the USSR sang praises of the nation's health and devoted an enormous amount of funds to developing medicine and sports.

In recent times, in the period of social states' turbulent development, medicine is starting to be analyzed as a socio-political institution developing and requiring enormous resources from the view of the effectiveness of using these resources to improve human health. One-factor determination of illnesses has proved erroneous even in cases of infections, to say nothing of chronic illnesses. It becomes clear that human adaptive possibilities are extremely great as long as the changes in the environment or in interactions with the environment do not proceed too rapidly.

In addition, medical aid can cause people harm associated with doctors' incompetence, the complexity and diversity of medical technology, and hypertrophy of diagnostic procedures. The institution of medicine is, like any bureaucratic organization, interested in self-preservation, and essentially achieves this through people frequently feeling ill, inventing diseases, and imposing the "patient role."

The reverse side of medicine's rapid growth is people's decreased ability for self-preservation and self-help. Medics harshly judge self-treatment, but this subcontracts the life and health of each person to specialists, disintegrates society, and violates autonomy and faith in oneself.

Expanding the sphere of medicine, which M. Foucault called society's "medicalization," is connected with the idea that a person is dependent on medicine from birth to death, while today an increasing number of specialists believe that ecosocial influences and a person's personal behavior are the main determining factors of illnesses.

The meaning of personal behavior and individual lifestyle for preserving health is generally underestimated in Russia.

Against the backdrop of "universal medicalization," there is a significantly widespread impression in society of older people as helpless, sickly subjects who are unable to make decisions independently or successfully fulfill social functions. This is associated with the fact that during the period of medicine's active institutionalization at the end of the nineteenth century, the first elderly patients were senile people with high rates of lost intellectual capacities. Based on studies of this group of old people, senior citizens were stereotyped, i.e. were considered to be a uniform group in which individual differences are erased and interests and problems are thought to be the same.

Older people were first referred to the office of medicine, although biologists and medics acknowledge that their professional approaches are entirely insufficient to define aging, since there are no cascading health declines in the majority of cases, and to be elderly does not mean to be ailing. It is imperative to change society's discriminatory attitudes, the discourse about "tallying diseases" formed by "yesterday's" medicine, and older people's fears that illness, weakness, and ailments are unavoidable in old age.

Today, the medical representations of functionality are legitimately secured and have unquestioned authority because older people are considered, first and foremost, to be medical and social aid subjects.

In Russia, these people are believed to be disadvantaged and in need of social protection and medical support... However, there are many reasons to believe that seniors' health problems are often connected with the increased stress load from rapid changes over the past 25 years (Roik 2011; Kozlova 2000).

We agree that seniors should be considered as any other legitimately differentiated social-age group, more so because its age limits begin at 55 years and continue through 90–100. Naturally, the elderly situation should be studied without emphasizing attention on illnesses and the need for medical help.

Meanwhile, more doctors are saying that representations of "cascading or linear health decline" with age are antiquated. A person can essentially remain entirely healthy, but why and how is not yet fully understood. There is need for "translational medicine," which brings knowledge about new achievements in "aging science" to physicians and the rest of the population.

Objective, multi-measure evaluation of senior citizen's health is crucial, considering the heterogeneousness of this social group. It is also important devote attention to the fact that the problem is not aging in itself, but the pathological type of aging that, unfortunately, is widespread in Russia.

5.2 Mass Media and Older Persons' Health

The subjective aspect defining older people's health problems is still connected with which problems visible to society, and which are not. In addition, according to N. Luhmann, it is almost as if "two realities" are formed (Luhmann 2008): the first is perceived by a specific subject and transmitted through information channels, the second is the reality perceived by information consumers. Successes and failures in the competition between social problems are not necessarily tied directly with the number of people effected by a problem. If the situation begins to be determined in mass media as a social problem, it does not necessarily signify worsening objective conditions. Likewise, a problem disappearing from informational space and social attention spheres, as S. Hilgartner and Ch. Bosk affirm, does not signify worsening position: "The fate... of problems is determined not just by their objective nature, but by a rigorous selection process, in which they vie against one another for social attention and societal resources."

Society can only see problems that are "shown and told" in mass media, as we live in a virtual world where everything depends on whether something or someone is discussed in conversation or writing. If something is not talked about, that is to say, media channels withhold some information, then the situation is not recognized in society and essentially stops existing for information recipients. Determining which topics exactly are repeated in mass media is also connected not to a problem's objective characteristics, but to understanding what the public will like, or in accordance with current political issues.

The dominating representation in society that old age and illness are inseparable exerts negative influence on the older people's status, because in associating with illness, old age is perceived as an undesirable state that disturbs an individual's physiological and social functionality and makes him or her dependent on others, ultimately leading to social exclusion. It follows to note that these representations are hopelessly antiquated. Today, unarguably, the majority of "elderly" diseases affect younger people, which suggests that they do not have an expressly age-related nature.

To preserve health, we emphasize the importance of prolonging older people's employment, since lengthening socially-oriented tasks, including paid occupation, is the "age longevity" concept's top priority. However, the media's effects are especially visible here, influencing belief that active longevity is linked primariliy with active care about personal health, or rather with elderly people's egocentric "obsession" on their own health, when in actuality the majority of older people report very low personal wellbeing satisfaction.

A. Smolkin notes that "in old age, a significant portion of events and actions begin to be measured or evaluated by a person in terms of health, and not satisfaction or need. Functionality itself as a norm basis depends on opportunities for medical intervention. Even social opinion tends to label dysfunctions that are amenable to change within medicine as normal" (Smolkin 2007).

Therefore, in the face of disease, an elderly person is scared not just by physiological effects, but by the change in relationships with surrounding people and exaggerated ideas about their own inferiority. These concerns can lead to decreased self-worth and nervosa. "Many older persons' actions associated with medicine are treated as disease symptoms, for example, the fact alone of visiting the hospital as a member of the third age is usually considered confirmation of his worsening health condition" (Smolkin 2007: 135). Unfortunately, representations about functionality in older age are medicalized, but comparative norms of functionality are established legislatively and psychologically-laden.

It is telling, that there are still no certain definitions of the concepts of health and disease in the system of knowledge, although they are key categories of medicine, healthcare, and even social work. The problem with understanding the nature and character of health and illness lies in the correlation between social and biological in a person. Health is considered today to be a condition as well as one of the final socio-economic development goals; for this reason, the concept of health should become a sociological category, i.e. be considered dependent on socio-cultural conditions. Society is inching closer to public health, but in general health is largely still a medical construction.

Health and sickness problems within the sociology of health and sickness occupy many classic sociological works of literature. "This area of learned knowledge includes a wide thematic spectrum: Hoffman's concept of stigmatization, E. Freidson's examination of healthcare's professional particularities, social attitudes toward the death (P. Hendel, V. Murphy), medicine as a social control institution (D. Tuckett)" (quote from Kovaleva 2001). This spectrum includes such concepts as Talcott Parsons' "role of the sick": society requires a certain level of health, which is a necessary condition for its existence. In his concept, health and illness problems are connected with preserving societal members' ability of to preserve their assigned roles. These abilities have both a physical and motivational aspect, which are closely interrelated, and because of which sickness is determined both biologically and socially.

Talcott Parsons aims this postulation at acknowledging the disease of deviations. Even such "unmotivated illnesses" as result from accidents, degenerative diseases, and infections contain a motivated aspect, as the individual can either consciously or unconsciously put him or herself at risk. Disease is dysfunctional for society and can be considered one of the ways of evading one's social responsibilities. Healthcare institutions, which include continuous reproduction of the roles of "doctor" and "patient," are needed to cope with disease.

According to a survey conducted by the Public Opinion Fund, two out of three Russian citizens do not pay attention to their personal health, at least when they are not sick, and this trend generally does not depend on education level, incomes, or social adaptation. However, the portion of people regularly doing something to promote health is somewhat higher amongst retirees (many of whose organisms simply won't let them forget about themselves) – 43%. In addition, proving highly characteristic of Russia, when talking about what threatens personal health most of all, most survey-takers selected factors that do not depend on themselves and that which they cannot change (43%) and only 7% selected bad habits and unhealthy lifestyle (Zhukov 2006: 66).

An opinion exists that the less people believe themselves responsible for their health, the higher the mortality rate. Of course, one can argue against the categoricalness of such formula, but in Russia, against the backdrop of high unnatural mortality before reaching retirement age, raising the population's level of responsibility for one's own health is extremely relevant.

Thus, we can simultaneously see widespread exaggeration of older peoples' sickliness and their need for medical aid, as well as the lack of reliable knowledge to assess their health and employment growth. This leads to a passive lifestyle and the unfounded fear that work after retirement age will lead to "residual health loss." Yet, it is necessary to assess this knowledge as well as the population's precepts in conditions of aging populations.

As an example, we highlight the importance of prolonging elderly employment to preserve their health as lengthening socially-oriented tasks, including paid employment, is a priority of the "active longevity" concept.

In addition, it is common knowledge that an active lifestyle is the key to a decently high and stable health level, because it slows development of involutional processes in an organism. Hence, "seeking medical help among working retirees is 6.1%, while non-working is 69.2% (Belokon 2006).

Furthermore, an enormous amount of statistical material is gathered in renowned gerontologist Vladimir N. Anisimov's works, and "analysis of the dynamics of the average age of death, starting from the first century B.C.E. through the end of the twentieth century, showed that average life expectancy for members of different professions gradually, but unevenly increased. Data collected confirms the viewpoint that high intellect and education directly correlate with high life expectancy and longevity" (Anisimov and Zharinov 2013).

But the media constantly imposes subjects and forms of consumption on the population, normative representations of health, sickness, and treatment methods, invoking demand in pharmacies for certain medications. Passing through the media, these representations gain unbelievable credibility and strength.

Studying Internet-messages showed that the topic of aging in general is surrounded by positive connotations. However, the joint phrase "health and old age" gives less optimistic results and shows that, on the one hand, the durable semantic bundle of aging and treating disease, but on the other hand, the presence of a significant number of diseases associated in people's consciousness with aging. In addition, these images and phrases are actively replicated in ads and promotional messages including on social networks, from the side of pharmaceutical as well as various medical and pseudo-medial companies selling products and services.

Older people reveal concentration on health and seeking medicine, and also seem to interpret active longevity as healthy longevity, oriented on seeking "medications or remedies for aging." Topics of mutual aid or active participation in social life are not yet discussed.

5.3 "The Role of the Elderly," "Secondary Benefits," and Adaptation Medicine

As mentioned above, the concept of the "patient role" was introduced by Talcott Parsons to designate a niche for an individual running from day-to-day stresses. This assumption lead Parsons to acknowledge the disease of deviations, since disease is dysfunctional for society, like a way of avoiding social obligations.

Parsons examines health as a "symbolic facilitator regulating human activeness and other life process as a valuable social product and substantial resource for individual achievements and smoothly-functioning society" (Parsons 1954). In addition, the patient status influences relationships with other social groups as well as society's relationship to the individual in general. Parsons believed the "existence of a number of institutionalized expectations, corresponding opinions, and sanctions" was proof of the "patient role's" presence.

Finding oneself in the "patient role," a person has the right, when applicable, to not fulfill usual social roles and remain blame-free due to illness; while also obligated to strive to heal as quickly as possible and follow the doctor's professional advice. The "patient role" provides a person with freedom from responsibilities for a time; "he is in a position where he must be cared for." Thus, acquiring the "patient's role" can lead either to the necessity to care for one's health, or to endeavor to maintain this role, because an individual in this position receives certain benefits and privileges.

The problem of preserving the Russian population's health is critical, especially concerning men, who have a low life expectancy and increasing mortality rate due to unnatural causes. As studies show, various factors substantially impact a person's health: social, political, economic, ecological, and psychological. Interactions between these factors strengthen their influence on one another, and in certain circumstances they can cause significant harm to health.

Indeed, "many of society's development trends are not oriented on supporting normal human organism functions. Unfavorable changes in health are frequently connected with low hygienic, valeological, and environmental literacy, as well as with behavior passivity" (Kovaleva 2001: 181). When talking about elders, one can say that in the current treatment situation, they face trends toward medicalizing old age and the uncertainty of the possibilities of biological control over it (Smolkin 2014).

Lifestyle is an extremely important factor influencing health at any age. This extends to a sustainable human way of life developed in specific socio-ecological conditions. By leading a certain lifestyle, an individual can affect irreparable damage or, in contrast, benefit to his or her health. This can be determined by both external factors and life approaches. Unfortunately, Russian retirees have not formed the skills needed for leading healthful lifestyles; in "caring" about one's own health they are used to passively relying on a healthcare institution, not on one's own efforts.

Visibly, they are used to understanding health as the condition of our organs, when they all "keep quiet," at least over certain periods of our life. So, according to O. Belokon's studies, "only 33% of old people try to obtain information about health problems at their age, and mainly elderly women display activeness in this. The fact requires attention that amongst the overwhelming majority, older people remark on a very low personal wellbeing satisfaction level. (Belokon 2006).

Many experts believe healthcare systems' influence on health risks to be less significant, but most Russian citizens value healthcare as the most important institution for tending after health. Public opinion surveys prove the fact that health invariably occupies one of the first places among life values, however it is true that if the question is open and any values can be named, few people mention health. The renowned sociologist L. Drobizheva attests that "Russians perceive health as an instrumental value: it is not needed by itself, but is needed because without it, it is impossible to attain more attractive things: good education, work, success with the opposite sex, and so forth" (Drobizheva 2004).

The "health" category is closely linked to the "adaptation" category. Renowned philosopher Ivan Illich, when noting the connection between adaptation and health, determined the latter as "the human ability to adapt to the undesirable genetic, climatic, chemical, or cultural consequences of economic development." Critically analyzing the authority of medical specialists, Illich also underlined the importance of confidence in oneself, one's abilities, and the capacity for fighting disease (Illich 1976).

In every country, depending on their socio-economic and national particularities, an infrastructure for medical aid and a system of forming a healthful lifestyle is developed and created, as well as control over its activities and the end result – the population's health. Here, it is impossible to ignore the role of self-preserving behavior as a health determinant. According several researchers, the main problem of self-preserving behavior consists of the discrepancy between consciousness and behavior. Frequently, individuals require health, however, it is not considered in relation to specific circumstances, i.e. the individual's need is not realized in his or her behavior.

"Self-preserving behavior includes attention to personal health, the ability to provide individual prevention for its breakdown, and conscious orientation on a healthful lifestyle" (Kovaleva 2001: 183–184). The most important determinants of a person's motivation for such behavior are the social values of health and healthy lifestyle, which must be fundamental and not instrumental.

Health should be perceived as a goal, not as a means. Thus, a person's behavior in relation to personal health is an important factor impacting physical and mental state.

Medical institutions that are interested in medicalizing old age and aging, i.e. the elderly, are a key contingent of medical service consumers, providing for existence and work by medical structures. In relation to this, "medicine becomes an institution of power, controlling peoples' lives and, in a number of cases, this control becomes unsupervised and acquires threatening forms," (Vlasova 2007: 188). Furthering this idea, it is important to remember of I. Illich's words, that modern medicine, in trying to strengthen influence over society, ceases to be aware of its limits and tries to replace religion.

Modern medicine's power over society is visible through imposing the idea that "only a doctor knows what disease is, who is sick, and what must be done. The meaning of patients' own complaints declines, as a result of which that person himself cannot determine if he is sick or healthy, but only the doctor can" (Plavinskij 2006: 33). Being an interested party, doctors are ready to declare any state an illness requiring treatment, which, in our opinion, is the manifestation of "governmentality," according to Foucault.

Therefore, society and, first and foremost senior citizens themselves, view themselves as victims of age, subject to one illness or another. True, older people inherently clothe any complaints in a somatized/temporal from, while doctors looks for the appropriate wording. Thus, the doctor and patient play the roles assigned to them in a social context. As a result, older people begin to think that any malaise is conditioned by a breakdown within the organism, and not as a result of a healthy unwillingness to adapt to difficult conditions surrounding life.

Thereby, paternalistic strategies of the state and medical institutions, on which the older generation was raised, facilitate rejecting an active life position and eliminate initiatives to overcome disease. Altaian sociologists' studies confirm this hypothesis. According to their research, 53.8% of surveyed older people believe that society's debt in relation to the old is support from all sides (material, moral), 38.9% state that society's duties include providing the elderly with a peaceful, problem-free, old age. This debt implies, first of all, creating the conditions for a comfortable life (Nevaeva 2014).

Not getting stuck on at which age society should support the elderly, we still express surprise at how it would be possible to support the "comfortable life" of one very numerous population group in a fragmented, uneasy society. And wouldn't advancing these stated demands add to the already high tension in intergenerational relationships?

But on the other hand, society itself sanctions the situation in which "you can turn to a medical institution with any problem, as countless popular medical literature, as well as ads for cures circulate images of diseases, indoctrinate ideas about the age-related unavoidable nature of diseases, and their fatal consequences" (Smolkin 2007: 139).

Many scholars note that today the lion's share of tumultuously growing medical expenses goes to diagnostic and treatment procedures, the effectiveness of which, at very least, is doubtable. This especially concerns all new pain medications, possessing extremely unpleasant side-effects. The surge of interest in Orthodoxy in our country has not yet led to accepting the most important Christian virtue – patience. In contrast, "patience and humility are entirely unwanted virtues, if only for the simple reason that they lower the incomes of pharmaceutical firms, which sell ever more expensive pain-relieving treatments," (Kosilova 2014).

Therefore, it is not worth exaggerating the abilities of modern medical technologies, as any interference can have negative effects, and pursuit of good health can lead to serious disorders and diseases. Escaping the vice-grip of medical dependence requires changing society's perception, and that of older people themselves: that in growing older a person is bound to face diseases, weakness, and ailments. Norwegian researchers Hagen and Moe collected proof over 20 years of work that age is not necessarily accompanied by worsening health. They conducted studies from 1987 through 2008 amongst elderly people over 67 in the affluent country Norway with its high quality of life, pensions, and excellent healthcare and social service system.

The data collected was compared with that other countries that are less prosperous in this area, and the researchers arrived at that very conclusion. The Norwegians noted an increased lifespan and a simultaneous decrease in the number of older people (including those over 80) with illnesses or disabilities: "The expected lifespan without functional limitation or disability grew even more for both sexes... Moreover, the proportion remaining in good health for life also grew" (Moe and Hagen 2011: 35).

Based on these population health studies, WHO experts note that older people often encounter special problems in terms of physical and mental health, which must be acknowledged, but that over recent decades, health has improved for members of the older generation (Mental, 2013).

Modern society, it seems, has figured out that not all elderly people end up senile. But now on the agenda is realistic assessment of the necessity for care and possibilities of both the state as well as the from the business sector in proposing inpatient and home-based services for elder care.

Increased medicine commercialization beats down mainly on retirees, whose financial abilities are very limited. There are additional problems, for example, older people living alone, isolated from medical organizations. In Russia, there are several tens of thousands of settlements, in which there are fewer than 100 inhabitants, and almost 25 thousand in which there are fewer than 10 inhabitants. In general, these people are elderly and old and are essentially devoid of access to medical aid as well as the opportunity for any kind of social adaptation.

Altaian University sociologists actively study elderly people's social adaptation, although they do not take account of health or access to medical services in social adaptation strategies. From their point of view, adaptation happens through socio-political activeness, economic behavior, migratory behavior, self-development/self-education, educational intentions, family-household interactions, and social interactions.

It is saddening to think that approximately 60% of both young and old survey-takers believe that norms developed over years and "life in one's world" are the sources of adaptation. It seems that people do not want to "sacrifice principles" and reject adaptation to changes in the real world. Sociologists conducted this research in Altai and Krasnoyarsk territories and the Kemerovo and Saratov regions (Maximova et al. 2017).

A. Kosilova sees a radically-ironic alternative to older peoples' concentration on diseases: "A positive alternative to the diseases of old age in culture serves the chief (appointed chief) characteristic of youth: sexuality. It would be possible to appoint something more "proper," but enough has already been said in our culture about wisdom. The cult of sexuality in modern culture in many ways serves the oblivion of death. Anti-aging remedies, whenever possible, emphasize returning namely to the sexy aspect of youth" (Kosilova 2014).

5.4 Death, Dying, and Loss in Old Age

The issue of death and dying in modern consciousness is suppressed, as a rule. It is considered indecent to talk about terminal illnesses, dying, and death. This relatively recently emerging stereotype of social consciousness and behavior is described in detail by Philippe Ariès in the fundamental study "Man in the Face of Death" (Ariès 1983). Experiencing internal uncertainty (an "existential crisis)" in the face of life and, especially in the face of death, is a relatively new phenomenon.

Also, in cases of unsuccessful grief flow, a person can displace everything he is connected to.

In the twentieth century, fear is developing toward facing death and even its very mention. It is true, however, that, it is against the backdrop of this fear and loneliness that people more often seek God, begin to read the bible, or go to church, i.e. return to the womb of Christian culture.

The tendency to displace death from the collective consciousness, gradually growing, reaches climax in our time when, according to Ariès, society behaves as if no one dies in general and an individual's death does not punch any holes in the fabric of society. In more industrialized Western countries, a person's death is furnished so that it becomes a matter only for doctors and funeral directors. Funerals are simpler and shorter, cremation has become a norm, yet mourning and crying over the deceased are perceived as some kind of "soul sickness". Death threatens the American "pursuit of happiness" as a misfortune and an obstacle, thus it is not just distanced from society's gaze, but hidden, and the modern "simulacrum of faith," psychology, is completely devoid of spiritual content.

Nevertheless, one of the most dramatic thoughts, on which every self-aware person strikes is becoming aware finitude of one's own existence or that of loved ones. According to the theories of several modern philosophers, when a being is in the final stages of life it turns to thoughts of death, finally, acquires genuineness, and becomes a form of authentic being, snatching a person from the realm of random possibilities and placing him in the womb of existence, where the person encounters life and self-understanding opportunities. In fact, finding meaning in life is connected to achieving self-understanding and valuing the uniqueness of individual "I" outside of social comparisons….

In many ways, involvement in Christian faith makes this self-understanding possible, at least in life's twilight. "He who transitions from this world to the next finds freedom." Studying the ethos of old age under the sign of the old testament, S. Lishaev points out that committing the discovery of the old testament elder ("the true old man," in K. Pigrov's terminology) freely and peacefully beholds the world, the middle, God, and opens himself willfully to good and prepares for the Crossing. This lifestyle is obviously connected with Christianity… (Lishaev 2010).

However, sometimes a person awaits death… "We are not so much talking about each person's right to death; suicide is not considered a crime in most countries. We are talking about patients suffering from terminal illnesses who, according to euthanasia supporters, are deprived of the opportunity to control their own life. Infringement on rights of this category of patients is noted in the situation of "intolerable suffering…" writes sociologist Elena Bogomyagkova (Bogomyagkova 2010).

Elderly people who sustain a sound mind to their last days, as a rule, resist placement in a hospital in their dying period, understanding that being is this impersonal, unfamiliar environment, not their House, will quicken their passing. The "garbage" metaphor originally issued the mouth of an old woman: "Just don't dump me in a hospital."

Unfortunately, in Russian society, where there is a fairly developed infrastructure for "welcoming into the world," there is almost an entire lack something similar, i.e. a symmetrical infrastructure for "sendoffs." As we have already mentioned, this is connected to the fear of death and the lack of faith or hope for other forms of life beyond this temporal and material one. Visibly, this is the biggest loss the modern person suffers due to lost faith, disbelief, and flat rationalism. As Yuri Gagarin supposedly said, he "didn't see God there [in outer space]." But if the soul does not exist, creating man in God's image, then there is no afterlife and no meeting outside of this mortal life.

Long-term illness in the final period of life makes an older person especially defenseless in relation to any external circumstances. Thus, instruments are being developed today for comparative study not of the "quality of life," or "happiness index," but of the opposite indicator, the "quality of death." In fact, "quality of death" is evaluated essentially as the quality of life in the Transition or Passing period; humane treatment at the very end of a person's earthly life.

Beginning in 2010, Singapore socio-environmental charity organization, The Lien Foundation, and the British company The Economist Intelligence Unit (an analytical subdivision of the Economist newspaper) conducted a study on "The Quality of Death." The study is named by its authors in analogy with the widely-known term "quality of life." Experts examined accessibility to social and medical services in select countries, the quality of these services, people's awareness that these services are accessible, as well as the particularities of national cultures associated with perception of death.

According to the study's results, Great Britain was named the most favorable country for dying people.

Setting up the ranking, scientists mastered the methodology of expert assessments, based on the conditions of 24 qualitative and quantitative indexes, integrated into four main groups:

– Basic social and medical services for people at the end of life.
– Availability and accessibility of services.

- Cost of services.
- Quality of services.

Great Britain's total score was 7.9, which significantly exceeds the indexes of many other European states, despite the fact that the British healthcare system is not considered the world's best. In the following order placed New Zealand, Belgium, Australia, the Netherlands, Germany, Canada, and the USA. Russia held 35th place in the ranking amongst 40 countries, settled between Turkey and Mexico. India closed the Index with 1.9 points.

In 2015, we saw a number of changes in the index's calculations. Russia became 48th out of 80 countries (which is higher than in 2010, as the earlier index considered 40, not 80 countries) by the Economist Intelligence Unit's rating, next to Turkey and Peru. The unchanged leaders were Great Britain, Australia, and New Zealand, while China was at the bottom of the list. In post-Soviet territories, Kazakhstan and Ukraine had worse results than Russia, and amongst the European Union, Slovakia, Greece, Romania, and Bulgaria.

When calculating the index, they now considered 20 indicators in five categories, which were assigned different values:

- Palliative medicine and healthcare (20%),
- Human resources (20%),
- Aid accessibility (20%)
- Aid quality (30%)
- Community involvement, including volunteers and families (10%)

The quality of death becomes a more critical issue as demand for a decent conclusion period of life grows. Russia, along with Canada, Portugal, Denmark, India, and several other countries, relates to the category of countries where, as assessed by the EIU, state strategy exists for developing pallitative medicine, but lacks clearly defined goals and has an advisory nature – regions are not required to strictly adhere to the pallitative care strategy.

Over the next few years, for the first time in human history the number of people over the age of 65 will exceed the number of children under the age of 5. The process of population aging is emerging particularly quickly in several developing countries, for example, China, due to disrupted generational balance. Today in China the small "middle generation" (born after 1980, when the "one family – one child" policy began) encounters the necessity to support the much larger Third and Fourth-Age generations. In addition, the pension system encompasses only the urban population, while residential care institutions were not accepted until recently.

Against the backdrop of a powerful demographic shift, the elderly portion of the population struggles with complicated and long-term illnesses of aging, which become more difficult and expensive to keep under control using family resources. In Russia, according to various assessments, 20–30% of families are multigenerational. Long-suffering elders become an enormous problem for older relatives. Notably, in a significant number of cases, care givers die before those they cared for.

In other cases, this leads to barely-matured children and grandchildren leaving family life early.

It is important to develop a system of palliative aid, but only in seven countries from the rating included this task in national policy in the field of healthcare and established it legislatively by 2010: Australia, Great Britain, Mexico, New Zealand, Poland, Turkey, and Switzerland.

Fortunately, the situation is changing, albeit slowly. In Russia in 2015, state subsidized access to palliative medicine programs with understandable criteria were proposed, however financing is not so easy to obtain and the programs' effectiveness are not tracked, as experts from the hospice charity fund "Vera" write (http://www.hospicefund.ru/fund/management/). One problem still remains: the low level of societal awareness on issues of palliative medicine, in particular, due to insufficient information on state websites.

A government's economic flourishment does not necessarily signify a better quality of care for its aging and terminally ill citizens. Research shows that many people in countries with a developed healthcare system also suffer from poor quality of death, even death from natural causes, as shown by authors of the presentation "Rating the Quality of Death in the Year 2015." In general, the quality of death becomes a more critical issue, while demand for a decent conclusion period of life grows (Quality Rating, 2015).

In the majority of countries, researchers have observed authority's aversion to dedicating attention to this issue, poorly-informed communities, and an insufficient level of preparation for medical personal caring for the aged and terminally ill. The situation is complicated by ambiguous relations to death and the cultural taboo associated with it.

Silence signifies that people are not psychologically or morally ready to experience Death, more so because it no longer seems to be the last test on the threshold of Eternity.

***

In this chapter, we wanted to understand what constitutes "elderly health." The result remains not fully convincing, as the health of elderly people is evaluated from the viewpoint of adults' health. It is difficult to imagine which distortions in assessment/diagnostics/treatment would arise from rejecting pediatrics and specific understanding in terms of children's health. However, for elderly people this happens in many ways, because discussion boils down to the number of older persons' apparent and hidden diseases....

We don't want to say that health is too serious a matter to entrust to doctors (paraphrasing the famous saying - war is too important a matter to entrust to military men), but once again we emphasize that responsibility of every person for looking after his or her own health, from infancy to the very last breath, is great.

In Russian media, there are almost no positive examples of productive aging. The only cases illuminated are insufficient, there is no large-scale campaign for leading an active and healthy lifestyle. The UN General Assembly highly values the contribution that older people make in the life of their society. The general declaration of human rights poses the following goal: to ensure implementation of universal norms in relation to specific population groups, in particular, to make a full life for older people, developing their abilities to self-regulate and self-organize, and to prolong active life as long as possible, increasing the period of self-sufficiency.

In conclusion, instead of unconstructively discussing the diseases and risks of aging, society should be more interested in wider development supporting medicine, mobile aid for seniors, and informing the population about technical opportunities and devices facilitating care. But people should also understand that there is no "pill for old age," especially against the backdrop of poor "self-care" at a younger age.

References

Algin, A. P. 1990. Risk: sushchnost, funkcii, determinaciya, raznovidnosti, metodologiya otsenki. *Socialno-filosofskij analiz*. Avtoref. dissertation. Moscow.

Anisimov, V.N., and G.M. Zharinov. 2013. Prodolzhitel'nost' zhizni i dolgozhitel'stvo u predstavitelej tvorcheskih professij. *Uspehi gerontologii* 26 (3): 405–416.

Ariès, F. 1983. *Images de l'homme devant la mort*. Paris: Seuil.

Belokon, O. 2006. Sovremennye problemy kachestva zhizni pozhilyh v Rossii (rezul'taty provedennyh oprosov). *Uspehi gerontologii* 17: 87–101.

Bogomyagkova, E. 2010. Evtanaziya kak social'naya problema: strategii problematizacii i deproblematizacii. *Zhurnal issledovanij social'noj politiki* 1: 33–52.

Drobizheva, L.M. 2004. Cennost zdorovya i kultura nezdorovya v Rossii. *Bezopasnost Evrazii* 1: 29–36.

Illich, I. 1976. *Limits to medicine; medical nemesis: The expropriation of health*. London: Marion Boyars Publishers.

Kosilova, E. 2014. O vzroslenii v multifaktornoi kulture. Otechestvennye zapiski .5. http://www.strana-oz.ru/2014/5/o-vzroslenii-v-multifaktornoy-kulture. Accessed 6 Dec 2017.

Kovaleva, N.G. 2001. Pozhilye lyudi: socialnoe samochuvstvie. *Sociologicheskie issledovaniya* 7: 73–79.

Kozlova, T.Z. 2000. Zdorovie pensionerov: samoocenka. *Sociologicheskie issledovaniya* 12: 89–93.

Lishaev, S.A. 2010. *Staroe i vethoe: Opyt filosofskogo istolkovaniya*. St. Petersburg: Aletejya.

Luhmann, N. 2008. *Law and states of exception*. Stuttgart: Lucius & Lucius.

Maximova, S., M. Maximov, and O. Noyanzina. 2017. Relation between civic attitudes, generalized and institutional trust in six regions of the Russian Federation. *Journal of management and markeng review* 2 (1): 24–32.

Merton, R.K. 1936. The unanticipated consequences of purposive social action. *American Sociological Review.* 1 (6): 894–904.

Moe, J.O., and T.P. Hagen. 2011. Trends and variation in mild disability and functional limitations among older adults in Norway, 1986–2008. *European Journal of Ageing* 3.

Nevaeva, D.A. 2014. Osobennosti socialnoj eksklyuzii lic pozhilogo vozrasta (po materialam sociologicheskogo oprosa). *Vestnik kemerovskogo gosudarstvennogo universiteta* 2 (58): 317.

Parsons, T. 1954. Values, motives, and systems of action. In *Toward a general theory of action*, ed. T. Parsons and E. Shils. Cambridge: Harvard University Press.

Plavinskij, S. 2006. Osoznala li medicina svoi predely. K 30-letiyu "Medicinskoj Nemezidy" Ajvana Illicha. *Otechestvennye zapiski* 3 (1): 23–38.

Rogozin, D.M. 2012a. Liberalizaciya stareniya, ili trud, znaniya i zdorov'e v starshem vozraste. *Sociologicheskij zhurnal* 4: 62–93.

————. 2012b. Pyat knig po liberalizacii stareniya. *Psihologiya zrelosti i stareniya* 4: 59–66.

Roik, V.D. 2011. *Mir pozhilyh lyudej i kak nam ego obustroit.* Moscow: Eksmo.

Smolkin, A.A. 2007. Medicinskij diskurs v konstruirovanii obraza starosti. *Zhurnal sociologii i socialnoj antropologii.* 10 (2): 134–141.

————. 2014. Vozrast: v poiskah sociologicheskoj optiki. *Sociologiya vlasti* 3: 7.

Vlasova O. 2007. Criteria for being Normal: Social phenomenology of madness. Zhurnal sociologii i socialnoy antropologii. Vol. 10. № 2: 184–190.

Yanickij, O.N. 2003. *Sociologiya riska.* Moscow.

Yatsemirskaya, R.S. 2006. *Lekcii po socialnoj gerontologii.* Moscow: Akademicheskij proekt.

Yatzemirskaya, R., and I. Belenkaya. 1999. *Social gerontology (Social'naya gerontologiya).* Moscow: VLADOS.

Zhukov, B. 2006. Russkiy sindrom. *Otechestvennye zapiski* 3 (1): 66.

Zhuravleva, I.V. 2006. *Otnoshenie k zdorovyu individa i obshchestva.* Moscow: Nauka.

Zubarevich, N. 2015. Regional dimension of the new Russian crisis. *Social Sciences.* 46 (4): 3–18.

Chapter 6
Social Service for the Elderly

Social service for elderly people is one of the most important focus areas for social policy, which social services can accomplish through various ownership forms: state, municipal, private, and public, etc. (Federalnyj zakon 2013). Despite significant changes in the legislature regulating social service, state social services currently predominate in Russia, but each of the RF's subjects has its own features or "Guaranteed List of Services," since these powers are transferred to RF subjects.

Liberalizing the concept of the state and its activity over the past 12–15 years has led to its gradual departure from many social activity spheres, to which the population was previously accustomed. However, neither a social insurance system that the population understands, nor complementary support for various forms of self-organization and mutual aid for the public, including within different communities, were created during this period. Publications on social policy note that the 1990s were a decade of change and revision in terms of many traditional social obligations in almost all developed countries (Hort 2004). Population aging served as the main reason for change, and opinions on the necessary social policy revisions in are found between increasing retirement age and decreasing pension size, i.e. insured compensation for lost earnings.

Western researchers have asserted that it is imperative to modify an entire set of institutions to satisfy the demand caused by society's changing age structure: to raise taxes on the working population in order to preserve older people's achieved level of life or to raise the retirement age to 67, and in the future – to 69 years (Caldwell et al. 2002).

In most of the world's developed countries, the retirement age was originally established at 65 years, while the planned survival time, i.e. life expectation at retirement, was 13 years. However, at present, in countries belonging to the Organization of Economic Co-operation and Development (OECD) the average expected lifespan of retired men is 17.9 years, while for women it is 22.8 years.

The Russian scientists' opinions also diverge: from the need to prepare for new challenges (Slepukhin and Yarskaya-Smirnova 2005) to assurance that the decreasing

© Springer Nature Switzerland AG 2019
I. Grigoryeva et al., *Elderly Population in Modern Russia*,
https://doi.org/10.1007/978-3-319-96619-9_6

number of children and, accordingly, expenditures connected with them will compensate for the increased expenses to support the elderly (Vishnevskij 2004).

6.1   Western Experience Serving the Elderly. Interaction of State and Community Support

The elderly person is social work's most traditional focus. Cruel and socially-excluding relationships towards older people, the roots of which were established in ancient times, unfortunately are not uncommon even today (Bocharov 2000: 169–184). Nevertheless, in the pre-industrial epoch when there was no retirement institution, lonely elders, having lost relatives and children for some reason, became social care subjects. Those people, because they were excluded from the most universal and traditional social care institution – the family – were often taken under the wing of communal and church organizations. Thus, in England, the "Law on the Poor," adopted by Queen Elizabeth in 1601, lay the basis for special legislature to regulate social charity. Special care was ministered to orphaned children and unattached elders, and they remained in the parish even if that meant placing them in the almshouse which, due to limited parish funds, hosted children, elders, and the ailing (Webb 1910: 21–23). Thus, old age, poverty, and family loss were not considered social exclusions, but instead all local inhabitants participated in preserving the community's wholeness, which both the church and the law supported.

In the present time, elderly social problems are also resolved using varied resources. The "economic mix" of social welfare often includes four main agents/subjects: the welfare state, the market (compulsory and voluntary social insurance), civil society (the volunteer sector), and informal associations (family, neighbors, friends) (Abrahamson 1993; Grigoryeva 1998). Civil society is a world of various associations, unions, and collectives, which has received the general and not very definitive name "community." Quite frequently family, neighborhoods, and local groups are also included within "community." The concept of the "third sector" offers a more precise definition: legally formed communal non-profit organizations possessing the status of non-governmental or non-commercial organizations (NGO/NCO). Meanwhile, the state is traditionally referred to as the first sector and economic structures as the second.

Industrialization disrupted the traditional family-neighbor commune, as a significant number of people began to work for hire outside of the family and local collective. In Europe, this led to the development of various social insurance forms, first for individual enterprises, then in the 1880s it formed in Germany, and later in other European countries, in a national-scale system, laying a foundation – the basis for the welfare state.

Pension insurance's advent gradually broke the connection between old age and poverty, since all people, first men and later women, received the opportunity to provision their old age through hired labor and compulsory insurance. Insurance, in our opinion, is the most perfect long-term mechanism of contractual arrangements between the generations of working people and retirees, where older people receive pension payouts in a general sum, introduced by the next generation working in accordance with the government's established rules and terms.

But insurance was not connected with permanent residence in some sort of local collective or family, but with individual labor relations. Pension insurance's development was a catalyst for elderly "washout" from the family, since each generation was now able to get along without the other. The necessity to have a lot of children disappeared so as to have someone to provide food and shelter in old age. The so-called family crisis consequently arose, as well as generational and gender independence growth. However, as a result of new independence, need emerged in social work (service) for compensational or supplemental family care.

In 1990–2000 in Europe, a rise of communitarian and solidarity ideology was noted, opposing state paternalism and rational insurance exchange relations. This phenomenon received the name "community development." Social work in communities was also called in English-speaking countries "community development," and in Germany it was known "communal social work." This social work objective was associated with broad local participation to resolve individual problems on a mutual benefit and cooperation basis. Social work acts here in the role of organizer, mastermind, and manager of this local neighborhood community, stimulating social self-organization. Moreover, elderly people specifically play the role of a community development resource, possessing the qualifications, life experience, and most importantly, time for resocialization and new social inclusion.

In the present, the market for "free time," treatment, and care is growing, as well as social service form diversity. In Europe, out of all people employed in this industry, approximately 22% on average work in the public sector, 70% for NGOs (Novyj obraz Karitas), and 5–10% are self-employed. If one was to comparing elderly social service organization in Great Britain, Sweden, and Russia, it becomes clear that, independent of historical traditions, the state (social bureaucracy) exits the service field in Europe and England. However, organizing this exit without large losses for the population is only possible by taking each country's cultural traditions into account.

In modern times, there are numerous initiatives and actions to benefit older people, which cannot be clearly attributed to the state, the social insurance system, nor to local government and civil society. They provide mixed, flexible ways of resolving social problems (welfare pluralism). Moreover, it is universally believed that even the best assisted living facilities or residual institutions cannot replace family and home stay. Accordingly, service should maintain an "open" nature, especially for the elderly (in sociology this is expressed via criticism of "total" or "closed" institutions, which are supervisory and effectively facilitate social exclusion).

It could be said that family, neighborhood, and local community have been rediscovered as basic "organic solidarity" institutions and social resource bearers. If the International Convention on Protecting Children's Rights accentuated a child's right to a family, the elderly population is now next in line. Mixed structures providing social policy in the form, for example, of volunteer organizations, cooperatives, or mutual aid/self-help for the population are no longer organizational innovations. In addition, they rely on the individual historical and cultural traditions in every country. One of these traditions is that elderly people are generally considered to be people ages 65 and older in many developed countries. Thus, in England, when speaking about social services for the elderly, this usually refers to people over 70 years old (Compilation... 1993).

All services are directed, primarily, toward providing for a family's normal existence for those who depend on the family most of all – children and the elderly. Today, in many European countries, there are no strict requirements for adult (grown-up) children to maintain and/or care for their parents. These costs are encompassed by traditional pension insurance and a new type of insurance: for long-term care as well as additional financing in need-based cases through federal and local budgets.

In France and Italy, supporting elders living at home with family has always been a priority, but this has now become an all-European trend. The belief is that service within family or one's home better supports elderly independence and is significantly cheaper than a retirement home. Thus, at the expense of the afore-mentioned insurance for final years' care, special aids have become available for home care. The insurance allows for either compensating the family member who cares for the older person for part the lost earnings, or for hiring a nursing aid. This is important, so as not to speed up transfer to a hospital. At home, groceries and meals can be delivered; special nurses/home aids/social workers help with bathing, laundry and changing undergarments, cutting and hair care, cleaning, and apartment repair and improvements. A special TV channel broadcasts Sunday mass for older Catholics who are unable to leave the house (attending mass on Sundays has long been more of a cultural than religious tradition; mass and Sunday lunch are universal family rituals).

The normal competition for social service quality helps the fact that they are at least partially free. Costs for paid services cannot be increased, and measures are taken so that those who pay more are richer. Good social marketing is when cost is competitive, but allows anyone to receive service. These factors make people not just passive users, but active clients whose demand for services determine an enterprise's success. Thus, in modern France, elderly service depends on increasing internal family resources, municipal social services, and NGOs/NCOs.

However, developing family resources is a difficult process. Renowned Swedish sociologist G. Espin-Anderson believes that modern families are included in a new type of economic relationships, the service economy, which regulates the family in an entirely new way (Espin-Andersen 2000). The service manufacturer faces up against housekeeping as an unbelievably strong adversary, because, in principle, the household can serve itself. Consequently, to broaden the service economy, it is essential to expand housekeeping demand to buy services or be insured "just in case." This might be the case if an older person living with a family lacks the strength and faculty for self-maintenance, while the family lacks time.

The new family has insufficient time particularly due to the revolution in women's motivations and preferences. That is why women, women's economic independence, and the social institution of family create a service economy and fill it with special logic. The new family type stimulates the new service economy, but also creates new social integration problems. It is possible that stability in family and community life is inversely proportional to the service demand level outside the home.

Swedish specialists, by the way, were the first to note that the family cannot withstand the competition of the various services for each of its members, so it is important to develop inviolable support for family community resources. Espin-Andersen believes that when family members are at work, in school, or day-care centers, local community thins out. Those who are free from work, for example, the unemployed or retired, face the neighborhood as the single possible means for inter-action, but these connections are unlikely to lead anywhere and their social exclusion could increase. Here his point of view contrasts with the English sociologists' faith in local community's integrational resources and availability of local communities' social capital, which can grow (Abrams et al. 1998).

The English authors stress that the form of elderly social support was restructured over the period from 1945 to the end of the 1990s. However, the society in which this reconstruction took place still carries several characteristics that have existed since the beginning of the nineteenth century. The transformations came from state centralized and standardized provision for fragmented support on the local level and in mixed economics conditions. This was a change in the provision method, but provision subjects/agencies – the state, the "volunteer" sector, the market and informal ties with family and friends – remained the same.

In England, the volunteer sector continued to play an important role in social support even during the period of the welfare state's rise. Particularly, during the period of state provision dominance, public groups formed to participate in solving problems that had been inadequately resolved. The importance of help for the elderly and increased poverty amongst retirees let to establishing such organizations as "Age Concern, "Help of Aged," and "Cinderella Services" (Symonds and Kelly 1998).

Members of self-help groups and political pressure groups reconstructed relationships between individuals, civil society, and the state. Hospital and social services were effectively transferred to local communities, despite economic growth and prosperity. Many authors diagnosed the situation as one of simultaneous "pressure from demographics, technology, and democratic equalization." Specifically, demographic pressure constituted population aging, when the number of people ages 75 and over noticeably grew and required increasingly more services.

The economic rise had ceased by the mid-1970s, and the predominating view of the state's position as the most effective services supplier was dismantled. Prime Minister M. Thatcher underscored the market's importance in driving changes and future provision. A new accent on individual responsibility emerged in terms of policy in the health field, and this movement expanded within the framework of individualistic discourse over the 1980s, accentuating overall attention to lifestyle.

Individualism was not a new phenomenon in British culture, but along with the individual and devotion to free market values, the concept of community "awoke,"

accompanying deep policy changes. Since in 1985, a movement began toward fragmented and mixed social security economics, placing the main responsibility to
serve the elderly on the community. The accent shifted from the state as the provision and service subject, to a state that authorizes, activates, and purchases services
from various suppliers. The state supported the growth of the volunteer sector's role.
Thus, M. Thatcher stated, "I believe that the volunteer movement is at the heart of
all our social welfare provision... The willingness of men and women to give service is one of freedom's greatest safeguards. It ensures that caring remains free from
political control. It leaves men and women independent enough to meet needs as
they see them, and not only as the State provides" (Thatcher 1976).

  In 1996, Prince Charles affirmed this commitment to tradition and entreated the
youth to devote time to volunteer activities, since that the problem remains that
primarily middle aged, middle class women participate in voluntary work. The
community's increased role and the financing with which Great Britain's government supports volunteer sector organizations have not escaped without difficulties.
Financing can proceed from many different sources, and this inconsistency can
summon resource dependence, curtailing the autonomy of organizations offering
service. Market values once flooded the altruistic world of volunteer help communities, which has had a profound effect on offering elderly people services.

  Social centers for all population groups have long been established in England
and the United States, organizing leisure and educational activities. These centers
usually operate from Monday through Sunday – 7 days a week. The working day is
from 8 to 23 hours and 60–80% of visitors are older people; these visitors are also
served by older people who are volunteer activists. Any person can come to a center
and begin to participate with a new group. This offers the opportunity to take part in
different types of activities, and any person can join for a small fee.

  The church and local schools offer these centers significant support. For example, the centers offer elderly supervision, so local schools organize concerts to collect funds and offer help, local shops offer coffee for morning care-givers and bags
of groceries for the elderly. By itself, the centers' work is based on broadening the
idea of "neighborliness" and mutual aid. The problem is that W. Beveridge, the
visionary behind England's "welfare state," suggested that women should work by
employment, not on housekeeping, only until marriage. Then they should turn their
attention to what we might call "work in one's place of residence," i.e. neighborly
mutual assistance, work in the church community and on school committees.
Women with small children and elderly women are to be the backbone of this work.
And although English women have long prior entered the labor market, some social
participation traditions have remained in the country.

  This is especially important for the elderly, since in many cases upon ceasing
labor activity the older person loses ties with the work collective to which he or she
was previously associated. Their children life independent lives, concerned with
their own problems. It is often the case that relatives live far away and do not have
the opportunity to visit the older person. "People think that old age is not accompanied by loneliness itself, but by the fear of it" (Pokrovsky 2008), which feeds into
the fear of aging and impending death. These are all stress factors on a person's

mind and nervous system, which can lead to a retiree's isolation from the surrounding world. A six-year study conducted by British researchers across more than 2000 people over the age of 50 showed that the mortality rate is twice as high among people living alone. The pain of being loneliness is comparable to physical pain. The organization Age UK talks about the necessity of combatting these phenomena and about local community's importance in overcoming this adversity (Loneliness... 2014).

The English "style" of working with the elderly contrasts with the Russian. In 2014, "the main part of the social service institutions – 7202 institutions – were under the regional government agencies' authority (RF subjects). Only 78 institutions or 1.1% were non-governmental. If state social service institutions offered services to more than 26.7 million people, then the non-governmental sector covered only 27 thousand citizens." (Vovchenko 2014).

In England, to the contrary, daily centers are generally municipal, rather than state-owned, although they can receive subsidies. The centers serve all inhabitants, primarily the elderly, by the "help yourself" principle, i.e. 2–3 people work on the center's payroll while the rest are volunteers. These centers have a very large "intake capacity," and do not have "shifts" or lists like we do in socio-leisure departments. Thus, while possibly cramped, the centers are not narrowly-specialized in either visitors' age or service areas.

It is very important, in our view, that the older person have the opportunity to bathe or receive hairdressing services at a daily visiting center; that it is a "white spot" in Saint Petersburg. These services for elderly people, which help them get from home to the center, i.e. with limited opportunities for movement, are especially needed. Daily centers should serve elderly people who are no longer able to get places; transport is needed not primarily for visiting museums, but to socialize with other people at least 3–4 days a week.

This draws attention the lack of state networking amongst English seniors, although we spoke with not-entirely-young people. We are not satisfied with care for aging women, when the most time is spent on volunteer work aimed at helping others, may be much older and or handicapped elders (we differentiate older people between the third and fourth ages). English elders more emotionally and warmly speak about their grown children, are proud of them, and support closer relationships with them, seeing them often rather than "very rarely." Women visiting the Specialized Subsidiary Office in Saint Petersburg and living in the country's cultural center, for some reason "necessitate" being taken to museums and having leisure activities organized. A very small number of women (traditionally, few men turn to Community Social Service Centers) talk about their desire to "give back," i.e. to help others. Possibly, this egoistic orientation could be corrected by social ads for volunteering.

Not just in Great Britain, but also in other European countries, these centers are financed by local budgets, private charities, and at the expense of care patients paying for services. On average, the fee could constitute 3 to 7 euros per day, depending on time spent at the center. Optional lunch could also exist as an additional charge of 1 euro. And primarily volunteers work in these centers.

In Germany, participating in social work gives the opportunity to receive a monthly non-taxable minimum of up to 500 euros, besides the fact that all expenses connected with volunteering are reimbursed. There is an old German saying that "a good Burger cannot sleep peacefully if there is an orphaned child or lonely elder who needs a piece of bread or roof overhead". Germans are meticulous taxpayers, and more than 60% of Germany's population fully agrees with the phrase: "I am responsible for how I feel in my country." German citizens acknowledge their responsibility for processes occuring in the country, choosing their active position. Experience working in the German non-commercial sector is impossible to fully copy and implement in Russian conditions, but it has many aspects worth considering.

"Insurance for care" is developing. Although our book is dedicated to older people of the third age, that does not mean we cannot talk about the needs of fourth age people and characteristics of their service. Today a consensus has been created in the world's developed countries concerning the necessity of radical reorganizing how long-term care services are offered. To raise a service's effectiveness, a substitution of fragmented help is implemented, provided by various institutions for more complex and coordinated long-term care involving various service providers. Focus is shifting from help in hospitals and retirement homes to extended services at home, for which a new insurance tariff is used, "treatement-aid."

Hence, for example, in Germany, mandatory insurance for long-term help and care (LTCI) was introduced in 1995 and is the fifth component of the social security system. Payments are established as a percent of income and are provided by workers and employers to varying extents. Insurance covers care at home and in specialized medical organizations, as well as financial stimulation, education, social and retirement provision for individuals offering care. Three coverage levels are envisioned, depending on specific needs and seriousness of the insured's condition. Care recipients bear financial responsibility for services, the cost of which exceeds the compensated payments, although social organizations pay for these services if their cost is too high.

Mandatory long-term care insurance (LTCI) in Japan also includes basic medical care, introduced in 2000 and financed on a parity basis by retirees (10–17%), the working population (33–45%), and the state (45–50%). Retirees pay a monthly amount determined by their salary (or their pension size), which varies depending on local jurisdiction or service type. People ages 40–65 pay a monthly determined fee along with employers in the social insurance system. The right to receive services is based exclusively on need and the system provides aid in associated institutions, as well as help for home care.

The nature of insurance coverage in different countries might be different (in terms of money and/or services). However, every country universally implements special programs to insure long-term elderly care.

Another interesting practice used by organizations helping the elderly in different countries are so-called specialized elderly settlements. Theses may be oriented at those who need care, as well as those who simply need socialization.

These communities are financed in different ways and offer varying services. They might be big or small, created with state help (in European countries) or without it (in the US). For example, "The Villages" in Florida consists of a variety of communities for active older people. In 2007, more than 75 thousand people lived in this settlement, and it was the most intensively-growing territory in the US. Most of the settlement maintains a "55+" format, meaning that only people over the age of 55 years can buy real estate there. The settlement "Cottesmore" in Southern California focuses on older people with weak health. It consists of private properties, a retirement home, and a restaurant. A wide spectrum of services are available to retirees, from help from a nurse/care-taker, to access to doctors with different specializations.

In Great Britain, a settled point was established under the name of Hartrigg Oaks. This settlement consists of 152 homes in which only older people live. Those who are unable to care for themselves can move into a large house where nurses can assist them. In this village, there is a mandatory inhabitance infrastructure. Village inhabitants pay an individual fee, and the entire community depends on the inflow of new residents. Meanwhile, in Japan, there is a special village for people with Alzheimer's disease. It is a living sector with a develop infrastructure, designed for people with impaired cognitive functions. Living in this village can help extend older people's independent life period.

In different countries, self-sufficient/independent life at home stipulates a medical examination and appointing services, which are executed by nurses, psychological support, personal physical care (offered by specially-trained workers) and mild housework, including feeding for an additional fee. To obtain these services, one must submit an application and complete a matching form. Upon receiving these documents, the social worker responsible for home care visits the applicant at home and conducts a need-based evaluation. Afterwards, the social worker submits the documents to a related council or committee for consideration, which makes the decision to offer service.

There are also services that stipulate selecting and installing additional equipment or fixtures for making life easier for older people, as well as providing instruction for using the equipment. This could entail installing support bars for the bath or toilet, or replacing the bath with a shower, adjusting furniture height, and a variety of fixtures which return a person the opportunity to be independent in carrying out daily functions.

Today, much attention is devoted to palliative care for older people. This is not surprising, since two thirds of cancer patients are people over 60 years, many of whom have experienced heart-attack, stroke, of suffer from Alzheimer's or Parkinson's diseases.

Palliative care is an approach allowing improved life quality for a patient facing life-threatening illnesses and his or her family, by mitigating an incurable disease's manifestations and preventing and lessening patient and family suffering. This is achieved thanks to early detection, careful examination, and curing physical symptoms, as well as providing psycho-social and spiritual support to the patient and his or her loved ones. In Great Britain, Canada, and Australia, palliative care is included in the educational programs for students within medical specialty and social work fields in their 3-4th year.

Forms of providing palliative care to patients are diverse; they diverge in different countries, since in each country these initiatives have developed along their own path. However, all palliative care forms can be divided into two main groups – help at home and inpatient.

Inpatient palliative care institutions include hospices, departments (wards) of palliative aid set on the base of general hospitals, cancer clinics, as well as stationary social protection institutions. Help at home is provided by field service specialists, organized as an individual structure or structural unit residential institutions.

Palliative care pursues the following goals: healing pain and other symptoms causing discomfort; affirming life and developing relation to death as a natural process (aiming neither to speed up or delay death's onset).

Palliative assistance includes psychological and spiritual patient care aspects; it offers patients a support system so that they can live actively for as long as possible until death; if offers a support system for a patient's loved ones during the time of illness, as well as during the bereavement period; it uses a multidisciplinary team approach to satisfying patients' needs and those of their relatives (during the bereavement period), increases the life quality, and positively influences the illness course.

***

A brief overview of changes in social service management in western countries shows that the transition from universal state social service system to selective addressing system based on different forms of mutual aid and service on a local level, is very complicated. It requires reliance on universal values, while perspective on these values is very different in Russia and in the West. Effectiveness calculations in Russia, where the insurance system still does not work, and relationships between the elderly and the state are described on the basis of understanding "debt," are entirely complex. Moreover, representing benefits and services in the natural form has the effect of self-selection, i.e. voluntarily rejecting/excluding those who are not too in need of this natural aid.

Transitioning again from the level of individualized needs to local communities' work, we note that international practice attests to the effectiveness of microfinancing for their development. This effectiveness is associated with the fact that the community itself pushes forth organizational leaders and decides for which needs request money. Naturally, independent problem resolution for the elderly with help of self-organization, social organizations, and modern informational resources requires time, patience, and motivation. But, as historical experience demonstrates, gradual changes are eventually more productive than the government's "wave of a magic wand."

6.2 Characteristics of the Elderly Service System in Russia

As mentioned in the introduction, a legal regulation has developed in Russia for social service of the population and the social services system. In 1995, two federal laws were enacted regulating the general principles providing social services and social care of older aged and handicapped people. From a formal legal point of view, these are not inferior to analogous western laws. They underline the necessity of providing equal opportunities to receive social services for older people, concentrating service on individual needs, and prioritizing measures of social adaptation for the elderly.

One can add that developing elderly social service is an important achievement for the government in recent years. During socialism pensions and inpatient clinics (retirement homes) existed, but with lost self-sufficiency amongst the older person who could no longer remain at home if not helped by relatives or friends. Moreover, such has been and still is the case that precisely these relatives initiated transition to the care home.

Today, however, lonely older people have the chance to receive social-daily services at home, generally twice a week (purchasing grocery products and medicine, helping prepare food, and helping clean the apartment, etc.) If the older person is not healthy, in addition to a social worker, a nurse might assist with daily visits.

There are social dining rooms and hair salons, offering natural assistance through products and clothes, etc. In many places, daily visit centers have opened where older people receive comprehensive service including such offerings as two meals and concerts. But service coverage was narrow, due to food and capacity limitation. To expand population service coverage, in the beginning of 2010, daily visit centers were created almost universally in social-leisure service departments, on one hand, and daily rehabilitation centers, on the other. Directing and coordinating activities for presenting social services are carried out by comprehensive public social learning centers, which currently exist in in every neighborhood center, and in every district in large cities.

Socio-economic transformations in the 1990s stipulated the necessity of an addressed/selective approach to older people's needs. But developing the addressed help system struggled against rejection by the documental confirmation system and regulatory checks both of client as well as social workers' needs. As with any change coming from above, addressed approach seemed unachievable not due to specific counteraction, but because "human material" inertia; commitment to the values of traditional culture, which characterize both professionals and clients.

Thus, at home service was constructed not based on character analysis and the client's self-sufficiency loss level, but based on the legally-accepted "List of Basic Social Services," i.e. perfectly standardized and impersonal, and thus identical for everyone. Naturally, the calculation did not account for neighbors or social organizations that could take on certain tasks, or purchase groceries. Even older people with a low level of lost self-reliance and who have children require that the social worker serves them, because "it is the duty of the state, for which we worked our entire lives," and so forth.

The goal of at home service is to create conditions in which any person can live self-sufficiently and independently in his or her accustomed social setting. Based on this definition, it is easy to examine the principles of self-reliance and mutual-aid, departing from socialistic paternalism towards communitarian ideology. In the West, it is believed that isolation and creating "special living places," even with the best intentions, firstly limit a person's rights and freedom, and secondly closes opportunities for development and life in "regular" society. Thus, even the most comfortable retirement home can be seen as a means of social exclusion.

Of course, this refers to elderly people who have maintained even the most minimal capacity for self-reliance and control over their personal situation. However, in Russia, these elderly people often believe otherwise, and there are those who strive to enter an inpatient center, since there "I will live all-inclusively; they will serve me."

On the other hand, in Russia and particularly in the provinces or "in the heartlands," there is a lot of undeveloped housing and older people have a hard time getting by, especially in the winter – heating the house, getting water, etc. The best practice, in this case, is to settle older people into temporary living sections in more-populated settlements. The standards of social service stipulate this opportunity, but, unfortunately, we do not know how often or universally this takes place.

However, there is a clear preference amongst elderly people for a dependent and passive position, allowing them to receive secondary benefits from their status as a victim of social transformations. Social passivity is accompanied by an older person's external identity focus, i.e. not considering the results of one's actions as the consequence of goal-oriented efforts, but rather attributing them to circumstance.

The inclination to place responsibility for what happens on external factors (the surrounding environment, fate, or accident) is characteristic of older people that, naturally, does not motivate the individual towards independent living or independent life in general. Yet it is impossible to deny definite progress in developing mutual aid for elderly people in Russia, as western researchers have noted (Harris 2011).

Nevertheless, A. Tocqueville stated that an observer detects state policy everywhere in France, in Britain there is the aristocrat's initiative, and in America one comes across voluntary association. Finding himself in Russia, he would undoubtedly notice that voluntary organizations play the role of friendship network. We could suggest that volunteering in relation to the elderly also needs to include a friendship network, which would develop amongst the elderly over time working in a labor cooperative and considering neighborhood connections (good-neighborliness), i.e. relationships by living place. This is all the more important because older people more rarely change their living places than other age group members and usually know each other for a long time. Volunteering will then have a more sustainable nature.

Over the past few years in Russia, the state policy foundations have formed thanks to mutual social-state efforts to promote volunteering support and development. Accordingly, the Russian Federation's Concept of Long-Term Socio-Economic Development in the Period up to the Year 2020 was created, affirmed by order of the Russian Federation government on November 17, 2008 No. 1662-p, to develop and distribute volunteer work related to the priority areas of social and youth policy. In 2009, the Russian Federation Government approved the Concept of Promoting the Development of Charitable Activities and Volunteering in the RF (Government 2009).

In modern Russia, the tradition of church aid has remained: in monasteries and churches there are almshouses that offer help to those who need it. The most famous in Saint Petersburg is the almshouse at Novodevichy monastery. Help is offered there to older women and weak sisters of the convent. Despite the fact that many have serious physical ailments, they help each other and serve the monastery's church. Of course, the monastery needs assistance to provide a more active life to its residents. Convent parishioners donate all possible means, which are used to purchase technical equipment and medical supplies for the charity people.

Borrowing from historical experience, retirement homes today are opening on monastery and church territories. Thus, the private Pokrovsky community retirement home in Saint Petersburg requires payment in the amount of approximately 50 thousand rubles a month, as of 2015. In this retirement home, there is a program to help elderly people who have lost housing due to fraudsters; they can live in the retirement home if enough help is received from charitable organizations (Pokrovskaya obitel 2009-2017).

Along with the Russian Orthodox churches, there are also different charitable organizations in Russia associated with other faiths. One of the oldest and most expansive Catholic organizations is "Caritas." It has representation in many countries worldwide, including in Russia. "Caritas goal is to motivate practical implementation social work, humanitarian aid, and human development amongst Catholics" (Novyj obraz Karitas 2002). Caritas works in many areas: with children, handicapped people, people with AIDS, the homeless, as well as with elderly people. In May of 2001 in Saint Petersburg, the "Caritas House for the Elderly and Lonely People" opened. The main task facing the institution's creators was to raise the level and quality of life amongst elderly people, particularly those who are alone. Aside from round-the-clock medical care, significant attention is devoted to recovering and developing self-reliance skills amongst senior citizens, and special rooms were created where older people can learn to live without someone else's help. The center offers psychological and social aid, since the older person's main problem is loneliness, the inability to share with someone and rely on another person.

The international Jewish organization Joint's work is conducted in Russia in three areas: helping senior citizens, supporting family and children, and developing Jewish communities. First in Saint Petersburg and later in other Russian cities, a unique system of Heseds was created (the word "Hesed" in Hebrew means "kindness" or "grace"). The system of Heseds develops centers offering a wide array of services to the elderly Jewish population based on the feeling of solidarity, "yiddishkeit."

These centers main taks is to "provide older people a dignified old age". The center works through many programs that offer aid in different fields:

- Material aid programs: "Winter Aid," "Emergency Assistance."
- Daily aid programs: "General Cleaning. Washing and Sealing Windows," "Meals on Wheels," "Grocery Home-Delivery."
- Medical aid programs: medical consultations, personal hygiene, "Vision Correction," "Medicine."

Within the frameworks of the Caritas and Hesed Avraham organizations, sustainable forms of volunteering have developed to benefit older people, which have been in the works for many generations already. Visibly, the Russian community needs to learn that the modern world is in an unstable equilibrium. According to Rosstat (Russian Federal State Statistics Service) data, in the year 2013, 113 thousand socially-oriented NCOs were tallied in Russia.

According to data from Rosstat studies in 2012, activity in education and science fields accounted for 25.4% of socially-oriented NCOs, social support and civil security for 21.9%, physical education and sports for 17.9%, patriotic and moral upbringing for children and youth for 14.7%, charity for 13.9%, healthcare for 10.9%, culture and are for 9.5%, offering various psychological aid for 9.4%, juridical for 8.8%, support for older people – 5.7%, support for the disabled – 5.5%, and for mothers and children – 4.2%.

This draws attention to the fact that only 5.7% of organizations supported the elderly, 5.5% the disabled, and 4.2% mothers and children. This is a small amount, but likely the 13.3% of charitable organizations and 21.9% of those working on "social support and civil security" can also be added to this field, as these things are interconnected.

In the present time question of substituting of the financial support for volunteering and third sector in general is relevant, since these were previously provided by Western funders. According to data from the Russian Ministry of Justice, in the year 2013, 2.7 thousand noncommercial organizations registered in Russia received foreign financing in the amount of 36 billion rubles.

Much has been done to replace the funds from western foundations. Thus, a competition for presidential grants was created in 2005. In 2014, the general monetary grant foundation constituted 2 billion and 698 million rubles, and 347 NCOs were awarded grants. In 2014, financial state support for civil society institutions in Russia consisted of approximately 10 billion rubles, and more than 3.2 billion rubles were allotted to accomplish socially significant projects (data from the Fifth Congress of Nonprofit Organizations in Russia, which took place December 2–5, 2014, in Moscow).

President V. V. Putin ordered provisioning NCOs with state support in the year 2015 in the amount of 4.2 billion rubles. Eight organizations acted as grant operators in 2015: the pan-Russian social movement "Civil Dignity" (528.5 million rubles), the pan-Russian social foundation "National Charitable Foundation" (585.6 million rubles), the Russian society "Knowledge" (695.6 million rubles), the Russian Youth Union (695.6 million rubles), the League of the Nation's Health

(519.9 million rubles), and the Institute of Socio-Economic and Political Studies (422.8 million rubles).

Two organizations distributed grants for the first time: The Russian Pensioner's Union and the Russian Women's Union.

The Russian Pensioner's Union directed 415 million rubles on projects to raise elderly people's life quality, social support for retirees, and social support for citizens in difficult life situations, offering assistance to those suffering as a result of disaster. This, of course, is very good, but it is not clear why the Youth Union was allotted 270 million more rubles, considering the situation that there are significantly fewer young people in the country than retirees.

For now, the volume of allotted funds is almost 10 times less than that coming to Russia's third sector from the West. For this reason, it is an important task for the state, regional leaders, and businesses to increase financing in order to preserve and support voluntary aid development for the elderly, which is especially relevant in the context of new social service legislation.

6.3 Elderly People as Mutual Aid Subjects

As the subjects of mutual aid and volunteer work, elderly people unavoidably encounter questions about their activity's politicization or refusing politicization. Volunteer initiatives in general can be considered collective actions, capable of forming socio-economic needs and obtaining changes from the government to better seniors' lives.

The new generation of older people in Russia first saw themselves as the "lost generation." They thought of themselves as victims of too-rapid socio-economic changes in the country. These were people who counted on retirement according to "soviet rules" and on receiving a high pension, although many years had passed since the USSR's decline. Possibly, they counted on continuing to work a little for a partial day/week, but thought that this would be by their choice. But they did not receive decent pensions, and understood that they would never again have permanent employment comparable to that which they had before retirement. And this large-numbered generation of "baby boomers" are 60–65 years old, i.e. a time when a person still has not lost anything in terms of qualification and work-capacity. As it turned out, the only work now suitable for older people is that which is unbefitting for the young: concierges, watchmen, and nannies, and etc. In other words, the social elevator only works "from below."

Another instance related to the fact that capitalism had arrived in Russia was that qualified older people, planning to work through the ages of 65–70, want not only to have a retirement age comparable to that in the West, but also a comparable salary. But the Russian doctor (in a state medical institution) or Russian university professor receives 3–5 times less in salary. Information about this is accessible via the Internet and fuels dissatisfaction among the older group.

Any collective action should be achieved with the help of some sort of coordinating mechanism, whether bureaucratic, network, or market, depending on the stability level of connection between the individuals acting within it. In the bureaucracy, interaction participants are determined based on formal rules, while in market interactions they are based on the equivalence of exchange between participants, and in the network they are personal acquaintance. The predominance of informal relations in network interaction in Russia quickly led not to self-organization, but to criminalizing part of society. Developing traditional self-organization, as once done in peasant Russia, is prevented by urbanization, the Internet, and open borders. If the chance of checking-out/exiting exists, it is much more difficult to force self-organization, because this is a collective matter – to find other people with whom to agree to do something.

People do not believe in the government because, even though there are elections there is a feeling that nothing changes. Thus, beginning in 2005 after adopting the law "On Monetizing Benefits…" retirees demanded to be returned no more than free public transit, but not to change the system!

It is believed that the horizontal faith deficit makes collective action impossible in general. Prior to industrialization and urbanization, there was a high level of faith at the level of family, kin, and neighbors, i.e. the local community. Now, political scientists believe that faith in the people you know has been transformed into faith in institutions. And a new culture is now emerging that comes from non-faith, but cultivating non-faith is a dangerous thing. If no one believes in anything, it is impossible to build up society. So attempts to create a Pensioners' Party and Pensioners' Professional Union slipped through the cracks. The Pensioners' Party was eventually created, but turned out to be incapable of playing any significant role in advancing decision making to improve older people's "quality of life" through their participation in NCO activities and independent senior organizations. True, now the Pensioners' Party is becoming one of the President's grants authorized operators, but this, in our opinion, speaks more about party functionaries' incomes than about senior self-organization (Zakon RF Ob osnovah… 1995).

Russia is similar with European countries in the level of elderly people (data from cities and city-like settlements) live separately from children and grandchildren at 70–80%. Aside from this, in Russia and in the West, the help often comes from older people to younger, rather than in the reverse direction. Charitable foundations are also not new to Russia. They emerged in the USSR prior to its collapse, successfully resolving complicated questions relating to socially-disadvantaged older people.

With the Internet's advent in Russia, older people's business and cultural-entertainment communication with their peers in any part of the globe became a daily reality. Serious work to resolve older people's problems is conducted by the Russian Academy of Medical Sciences' Interdepartmental Scientific Council on Gerontology and Geriatrics along with ministries of the RF. Large-scale gerontological centers are emerging and developing in Saint Petersburg, Samara, Ulyanovsk, Yaroslavl, Surgut, and other cities.

However, we note that the country's aging population problem is primarily managed by doctors and social security services (i.e. social services), which cannot presently be considered sufficient. A doctor can heal an ailment, but is not in a position to help a person settle into society. The social security system also cannot manage an aging population's negative social consequences (Kuvshinova 2009). The solution might be through joint work of services of employment and social service, but this is not currently in practice.

Thus, social initiatives are updated and social unions' activity become relevant, activating citizens and summoning them to participate in solving gerontological problems. Society's opportunities using older people's potential are far from exhausted. Little is done to decrease the level of social tension and age discrimination in society, despite the fact that in Russia, according to western authors, "ageism is combined in different ways with "traditional gender." And low pension, and alcoholism amongst men, leading to their decreased lifespan, increase the danger of social exclusion for women, which they conquer not so much with the help of social services, as through their own networks of acquaintances and neighborly support" (Jappinen et al. 2011).

At the same time, in Russia only 3.02% of the economically active population is involved in volunteer work of non-commercial sector organizations (Potencial 2010), while approximately half of citizens, according to various data sources, express readiness to participate in activity on a voluntary basis. However, here, as in the case of many surveys, people likely respond as "they should." The most popular type of volunteer work for Russians is planting flowers, trees, and lawns, and landscaping, and 27% of survey-takers are prepared to take part in this, while a lesser portion of those surveyed were willing to assist the elderly. Visibly, social work specialists specially will need to work on motivating the population towards socially-focused activity.

We refer to studying the social policy classic by English sociologist R. Titmuss (Titmuss 1970). Almost half a century ago he stated: if we start to pay for giving blood, the number of donors will decrease. Economists reacted to the scientist's observations with disbelief and accused him of lack of proof. However, new data has forced them to drastically reconsider their views on stimuli – in particular, it became clear that the "Titmuss Effect" occurs fairly often, including in business.

Experiments have demonstrated that if women are offered money to give blood, the number of donors amongst them decreases by almost half. However, if they are then asked to donate this money to charity, the prejudice is lifted and the number of volunteers returns to its previous proportion.

Benefit from donation drives other destructive forces. Related moral values are by and large undermined, particularly the important aspects of honesty and trust. Titmuss noticed that in comparison with voluntary donors, the statistical average paid donor "is less willing and less likely to disclose the full picture of health. From fear of being recognized as an unsuitable donor, he begins to hide information about recent contact with infectious patients, habits, nutrition, alcohol and drug use" (Titmuss 1970: 151). Accordingly, blood donated on a commercial basis is more likely to be infected and unsuitable for transfusion.

Russian practice in recent years has confirmed that financial stimulation is not necessarily effective and can lead to a contradictory result, undermining what A. Smith called "moral sense" (Bowles 2009). Completing certain actions, buying things, or trying to get a job – we do not want this to go against our beliefs. If this happens in the process of commercializing "everything and anything," then later "moral decline" happens very quickly.

Human relations are replaced by "dummy goods," according to K. Polanyi, i.e. goods which by nature and moral reasons were originally not goods but the effects of developing economic relationship and trends in prevailing in our social world become such: "A kind of disembedding which spawned claims from the side of economic relations on absolute superiority... over society began to be represented as something developed from economics" (Polani 1993: 32).

As a result, "commodification" of social services can develop within certain limits, without disrupting normal human relationships and the desire to help one another. Visibly, paid services should develop where qualified help is needed or the ability for self-reliance of an older person is already minimal.

Discussion about "market society" has long been conducted in the West. The interrelations of the "systemic world (government and market)" with the suppressed "life world (human and society)" are extremely important. German sociologists and, possibly, first Jürgen Habermas, demonstrated at the end of the 1970s, how excessive care of the state not only disrupts necessary solidarity bonds of the "life world," but also forms dependent clients amongst citizens. People, who for a long-term are not working due to large social benefits which, as clarified, have a "demotivating" nature, stop being citizens even "once in four years," forming the state's needed obedient electorate.

Despite the low benefit and pension level in Russia, social service clients also quickly become consumers, using the existing system's advantages. As in the West, social service clients' relationships to the government have a contradictory nature. Alongside discontent with poor care, people constantly await some kind of improvement in their position from the government's side, and the state in turn constantly emphasizes that it "will not let down" its main electorate....

Nevertheless, partially due to state order in the early 2000s concerned with motivating civil society and volunteering, nongovernmental and noncommercial organizations are developing in Russia, serving the elderly. Their role in connection with social service legislature updated in 2015 is also growing.

Select "Socially-oriented NGOs" in the search engine "Yandex" to find a "NCO Catalogue" and "Practices Search." Querying about client categories "Clients: Pensioners and Elderly People" comes in real-time approximately 200 NCOs which work on primarily identical types of aid. In addition, within the selection there are organizations of the society "Knowledge," and "Consumer Society," and an entire set of other NCOs which are entirely not specialized for working with retirees or the elderly. One also observes that there are many organizations within the European half of the RF, but few beyond the Urals.

The sources of financing among most of the organizations are the same:

- Russian organization grants
- Donations from enterprises and organizations
- Donations from individuals
- Providing paid services
- Membership fees, rare among other sources.

In terms of new interesting working organizations with active young directors, one can notice the Regional public organization of the Republic of Bashkortostan's "Coordination and resource center for seniors "My years are my wealth". Unfortunately, the information on its website differs little from other Social Oriented (SO) NCOs, although "My Years are My Wealth" is an option for "Volunteer projects aimed at the elderly offering experienced help to children and teenagers." These projects are: "Grandmas to children"; "75 and over – Life is Just Beginning!" and "Computer Literacy and Group Interaction – support for grandmothers raising orphaned grandchildren."

This is very important – to develop older people's subjectivity, support their ability and desire to give back spiritual warmth and aid other people, and NCOs support this. As already mentioned, implementing social programs in any organization in Western Europe and Northern America is accomplished by a small number of employees and many volunteers from different communities, religious, charitable, youth, and other organizations. In Northern Europe, a significant percentage of volunteers belong to the older generation. In Russia, elderly people are firstly ascribed the role of subjects – help-recipients, who themselves do not feel they are discriminated against.

In post-Soviet countries, elderly people's position is often worse than in Russia. For example, in Kyrgyzstan, older people are not only poor but often alone, since their children have left for work or gone permanently back to Russia or farther. However, their self-organization level in the context of minimal state help is fairly high.

Since the 1990s an NGO has existed called "The Resource Center for the Elderly" (RCE) – this is a non-governmental non-commercial organization working since 1991 in the field of social support and security for Kyrgyzstan's elderly population. The RCE's tasks include helping improve the system providing access to basic social services, drawing society attention to older people's problems, and developing and implementing social standards and services activities for organizations and institutions, taking international experience into account.

In the process of forming new public-state institutions and relationships, Russian authorities are required to answer new summons. In addition, the bureaucracy considers "civil society" not as a social institution, but as an instrument for realizing the political lines of one or another interested actor.

Thus, in the conditions of decreased political competition and completed transition from political system to an actual one-party system, some measures or proposals initiated by civil society structures were initially treated as dysfunctional and everything done so as not to allow them into normal discussion. Accordingly, civil society was developed with almost full financial dependence on President's grants.

Although modernization is impossible without changes, in modern Russia these changes are "determined" political in the framework of conservative discourse. In the Russian modernization model, authority acts as the initiator and organizer; the modernization goals are determined to benefit the elite; authorities' goals and funds do not correspond with the trend in the population's needs, which feels as if it has been "banished to another planet."

A normative model of older people's relationships with society still has not been found. It is believed that elderly people have difficulty accepting social and cultural changes and end up amongst the number of vulnerable social groups. In many cases, older people and their relatives do not know where to turn with their problems, which speaks not so much about their retrograde conditions, as about poor social service work informing the population about service opportunities.

6.4   Conclusion

Despite certain and even significant moments by which the elderly social service system in Russia can be criticized, over the past 25 years it has developed essentially from the ground up. Its main defect is in perceiving elderly people as passive subjects to whom the government "offers kindness." Overcoming this relationship is only possible via older people's participation in mutual aid, serving one another, or serving the extremely old. The main task is not to save budget money, but to increase coherence, society cohesion, and give people a way to "see each other."

Back in 2003, the first Russian geriatric center's organizer (Saint Petersburg), dying before her time, E. S. Puskova noted: "... the existing (social service) system is directed exclusively to make the desired be accepted as the actual. The city is covered, for example, by a network of at-home social services. But when a person, a social worker, comes and brings groceries, that is not social service in the sense for which this department was created. In addition, this person, for example, cannot say that he bathed the older person or cooked there... so it was not like some extra burden, but completely natural. Offering certain sanitary services, connection with relatives, talking, impacting family members – our social workers do all of this."

Unfortunately, this is all the more relevant today. The superficial and episodic nature of attention to seniors both in comprehensive public social learning centers and NCOs (holidays, presents, greeting cards, commemorations, tea times, and festivals, etc.) long ago became a "byline." We do not mean delivering groceries, which shopping networks could do perfectly well on their own, but specifically attention and care, which should have a continuous nature. In other words, businesses and shopping centers could offer some services, and the NCO could offer others.

However, interruptions in third sector financing indicate that sustainable activity models with senior participation and stable service assortment for older people are still lacking. The exception, it would seem, is again Hesed Avraham in Saint

Petersburg, where training is regularly conducted for both employees and volunteers. Visibly, a certain level of ageism is characteristic amongst NCOs as well, whose directors feel that additional qualification is not needed to work with senior citizens.

Aging is a triumph of development, as affirmed in the UN presentation (Bussolo et al. 2015). Increased lifespan is one of humankind's greatest achievements. The life expectancy at birth now exceeds 80 years in 33 countries; only five years ago this was the index for only 19 countries. Many people who read this report will live to be 80, 90, and even 100 years old. Even in Russia, people who were born in the 2000s have the chance to live, on average, to the age of 100.

We emphasize for those who too biased to see this process, in many cases and different countries people of an older age help their children and grandchildren, taking responsibility not only for raising children and housework, but also providing the family significant financial means.

It is necessary to provide all layers of society with different opportunities to receive education, work, medical service, and basic social services that allow older people to lead a decent lifestyle and working people to establish savings for the future. Especially in modern conditions, when the state seemingly knew that elderly social service attains an entirely bureaucratic and soulless view without volunteer participation, it is necessary to raise the qualifications of those who work with it, from NCO leaders to program organizers and volunteers directly.

It is imperative to develop non-stationary assistance for seniors and their families, informing the population about technical care opportunities easing independent living at home. But acceptable institutional accompaniment for such a "caring" family or elderly daughter, required to abandon work to constantly be near a seriously ill relative, is still in the making.

For now, the media exploits the topic of elderly care as a moral debt, not delving into the daily requirements and the difficulties this care entails. And gender sociology found a new force application point in which characteristic women's liberation motives intricately woven from "domestic slavery" and concern about the excessive involvement of women in situations of caring for elders and the "sandwich generation" overload.

Society and state should, at last, learn to see older people with respect and develop long-term development programs, the priority of which should be cultivating elderly participation in all social life spheres.

References

Abrahamson, P. 1993. *Social policy in changing Europe*. Roskilde: Roskilde University Press.

Abrams, Ph., R. Snaith, et al. 1998. *Community care: A reader. Neighbourhood care and social policy*. London.

Bocharov, V.V. 2000. *Antropologiya vozrasta*. St. Petersburg: Izd-vo S.-Peterb. un-ta.

Bowles, S. 2009. Russian: Materialnye stimuly: obratnaya reakciya. http://hbr-russia.ru/upravlenie/motivatsiya/a9840/#ixzz2y2SjOgdT. Accessed 6 Dec 2017.

Bussolo, M., J. Koettl, and E. Sinnott, eds. 2015. *Golden Aging. Prospects for healthy, active, and prosperous aging in europe and central asia. International bank for reconstruction and development.* New York: The World Bank.

Caldwell, J.C., P. Caldwell, and P. McDonald. 2002. Policy responses to low fertility and its consequences: A global survey. *Journal of Population Research* 19 (1): 1–24. May.

Compilation of social security laws. n.d. Including the social security act as amended and related enactments through Jan. 1. 1993. Vol. 1.

Government of the Russian Federation. 2009. Koncepcija sodejstvija razvitiju blagotvoritel'noy dejatel'nosti i dobrovol'chestva v Rossijskoj Federacii: Utverzhdena Rasporjazheniem Pravitel'stva Rossijskoj Federacii № 1054-rp ot 30 iyunya 2009 (A concept to facilitate the development of charitable activities and volunteering in the Russian Federation. Approved by Federal Government Decree № 1054, June 30, 2009). Moscow: Government of the Russian Federation. http://www.economy.gov.ru/minec/activity/sections/admReform/publicsociety/doc091224_1949. Accessed 6 Dec 2017.

Espin-Andersen, G. 2000. Two societies, one sociology, and no theory. *British Journal of Sociology* 51 (1): 59–77. January-March.

Federalnyj zakon Rossijskoj Federacii ot 28 dekabrya 2013 g. No 442-FZ «Ob osnovah socialnogo obsluzhivaniya grazhdan v Rossijskoj Federacii». 2013. *Rossijskaya gazeta* 295.

Grigoreva, I.A. 1998. *Socialnaya politika i socialnoe reformirovanie v Rossii v 90-h godah.* St. Petersburg: Obrazovanie-Kultura.

Harris, J.G. 2011. Serving the elderly: Informal care networks and formal social services in St.-Petersburg. In *Gazing at welfare, gender and agency in post-socialist countries,* ed. Maija Jappinen, Meri Kulmala, and Aino Saarinen. Cambridge: Cambridge Scholars Publishing.

Hort, S.E.O. 2004. Western experience–eastern experiment? Welfare state model and theories before and after the "Decennium Horribile". In *Sb. Statey,* ed. I. Grigoryevoy, N. Kildal, C. Kyunle, and V. Mnina. St. Petersburg: Skifiya-Print.

Jappinen, Maija, Meri Kulmala, and Aino Saarinen. 2011. *Gazing at welfare, gender and agency in post-socialist countries.* Cambridge: Cambridge Scholars Publishing.

Kuvshinova, O. A. 2009. Socialnaya reabilitaciya lits pensionnogo vozrasta: problemy i perspektivy. *Voprosy upravleniya* 8. http://vestnik.uapa.ru/ru/issue/2009/03/10/ Accessed 6 Dec 2017.

Loneliness twice as unhealthy as obesity for older people, study finds. 2014. *The Guardian.* http://www.theguardian.com/science/2014/feb/16/loneliness-twice-as-unhealthy-as-obesity-older-people/. Accessed 6 Dec 2017.

Novyj obraz Karitas: Kratkij obzor deyatelnosti «Karitas» v Rossii za period 1999–2001 gg. 2002. SPb.

Pokrovskaya obitel. *Dom prestarelyh.* 2009.–2017. http://www.omophor.ru/projects/nursing-home-pokrovsky-obitel. Accessed 6 Dec 2017.

Pokrovsky, N. E. 2008. Universum odinochestva: sociologicheskie i psihologicheskie ocherki. M.: Universitetskaya kniga, Logos.

Polani, K. 1993. Samoreguliruyushchijsya rynok i fiktivnye tovary: trud, zemlya i dengi. *THESIS* 2: 10–17.

Potencial i puti razvitiya filantropii v Rossii. 2010. Eds. Mersiyanova I. V., and L. I. Yakobson. Moscow: Izd. dom Gos. un-ta. vysshej shkoly ehkonomki.

Slepuhin, A.Y., and E.R. Yarskaya-Smirnova. 2005. Socialnye gosudarstva pered novymi vyzovami. Socialnye processy i socialnye otnosheniya v sovremennoj Rossii. In *IV Mezhdunarodnyj socialnyj congress,* 338–340. Moscow: RGSU.

Symonds, A., and A. Kelly, eds. 1998. *The social construction of community care.* London: Macmillan.

Thatcher, M. 1976. Speech to social services conference dinner. ("The Healthy Society"). https://www.margaretthatcher.org/document/103161

Titmuss, R.M. 1970. The gift relationship. In *From human blood to social policy*. London: London School of Economics Books.

Vishnevskij, A.G. 2004. Pyat vyzovov novogo veka. *Mir Rossii* 2: 3–12.

Vovchenko, A. 2014. *Only 78 of the population's social service institutions are non-governmental*. http://www.rosmintrud.ru/social/service/67. Accessed 6 Dec 2017.

Webb, B. 1910. *English poor law policy*. London: Longmans Green.

Zakon, RF. 1995. *Ob osnovah socialnogo obsluzhivaniya naseleniya v Rossijskoj Federacii*. No 195-FZ ot 10.12.

Chapter 7
Education Ideas and ICT Training Practices for Older Persons

Interpreting inclusion/exclusion processes in modern society is closely tied to changes in understanding how society's social structure is formed, how social mobility is constructed, and which resources are divided amongst members of different groups.

The traditional understanding of social structure as vertical has been supplemented by new factors overtime, indicating more than just individual's economic wellbeing/deprivation. In recent decades, the definition of society's structure has been transformed, becoming a rounded representation of space and having both vertical and horizontal measures.

With development of consumer society and globalization processes, classics modern sociology literature (Bourdieu 2005; Bauman 2004; Giddens 1990) outlined the appearance of new, numerous socially-excluded groups, tossed to the wayside of social life. In the parallel-developing concept of "informational-communicational-network" exclusion is associated with society faction's inability to access technological novelties, on the one hand, and absence of constantly-updated knowledge for their use, on the other (Vershinskaya 2015; Castells 2009).

Global population aging, becoming a critical agenda topic especially in developed European countries (Technical Innovations…2012; Riva et al. 2014), updated the issue of social exclusion associated with limited knowledge and resources for fully including of older people in the modern world of information technology.

The international "Okinawa Charter" advises "devoting special attention to the needs and abilities of people who have less social protection, people with limited ability, as well as senior citizens, and actively take measures aimed at providing them with easier access to the world of ICT" (The Okinawa Charter 2000). At the level of day-to-day life, the rapid loss of ties between young and older relatives signaled the substantial group of informationally-excluded elders' appearance.

Distributing accents in this way is more relevant to Western countries, where the retirement age begins significantly later than in Russia. In Russian, however, as pointed out above in Chap. 1, the problem of "older" social exclusion is connected

© Springer Nature Switzerland AG 2019   117
I. Grigoryeva et al., *Elderly Population in Modern Russia*,
https://doi.org/10.1007/978-3-319-96619-9_7

with their informational distance from the "young," as well as with early retirement, disinterest in work activity and/or no desire to prolong one's employment, and lacking skills befitting labor market demand.

One can talk about representations rooted in the social consciousness about the elderly person as handicapped in terms of innovation, assimilating/acquiring new information, and preparedness for learning and relearning. "Pressing social stereotypes build up such strength," M. Elutina and E. Chekanova note, "that the majority of older people construct their behavior in accordance with this label, creating thusly their own "handicap," which is incorporated into internal structure of identity, becoming a barrier for its self-realization. At the age of fifty, working professionals consider their careers to be over. Thus, they begin to prepare for retirement" (Elutina and Chekanova 2003).

According to the results of other studies, elderly people are not perceived more negatively than younger people (Krasnova 1998; Smirnova 2008). But according to A. Pisarev, one fourth of survey respondents believe that this social group is a deterrent to social development, while 78% sees elderly people as vulnerable and in need of social aid. However, 60% view see the potential in this group, which should be coaxed into a socially active life" (Pisarev 2004).

The existing array of stereotypes in relation to elderly people in Russia have not changed over the past 20–25 years: "In the social consciousness, a stereotype has firmly taking root perceiving this age group that, in general, propogates gerontophobic beliefs in relation to the period of old age and is projected not just by society, but also by the elderly people themselves in relation to their own aging" (Butueva 2014; Kovaleva 2008).

Our behavior, changing with age, is built on perceived meanings of what society considers to be appropriate youthful, adult, or "old person" behavior. But the actual situation is significantly more multifarious, and differences between ages lose sense if assuming the normalcy of a nonlinear life path.

However, the "normative model" of elderly interaction with modern society is linked to active inclusion in social life, including fairly lengthy employment (Grigoryeva and others, 2014a, b).

It's assumed that the absence of skills, in particular, for using information-communication technologies (ICT) is one of the main reasons elderly people become socially excluded from modern society (Vershinkaya 2015; Saponov and Smolkin 2012; Grigoryeva and Chernyshova 2009).

In the recent past, people have discussed not informational inequality or informational poverty (a UN Development Program 1997), but about a digital divide. To maintain older people's social inclusion, the government has recently unveiled a widely-accessible computer technologies training program in social services and has started refurbishing regional libraries.

The problem of elderly education was originally posed at the beginning of the twentieth century. In the words of Tatiana Kononygina, "researchers note two development directions for this process. The first was connected with the necessity of training those who work with elderly people and prepare them for a disease-free transition to old age. The second assumed elderly peoples' participation in education programs and projects with the goal of satisfying cognitive requirements, as well as a well-rounded preparation for an active social life after retirement.

Thus, the teaching discipline "geragogics" or "gerontologics" were formed as part of age pedagogy, a sphere of pedagogical gerontology, or as an area of interventional gerontology" (Kononygina 2006).

In our study, we decided to concentrate on the real requirements to satisfy older people, coming to study ICT, as well as learn what those who teach computer and Internet work skills think about this subject.

7.1 Research Methodology

The purpose of this study as a whole (from 2014–2016) consisted of evaluating the influence of new interaction technologies between elderly people and society (in particular, computer literacy training) on social inclusion. In terms of technologies' interaction, we consider elderly social service that is legislatively and institutionally affixed, and is used by government social services working with older people.

This chapter features the results of in-depth interviews with older people taking computer literacy courses at governmental social service centers, libraries, and specialized NGOs serving the elderly population in Saint Petersburg and surrounding regions. In total, we collected 30 interviews.

The main hypothesis of this part of the study was that elderly people's motivation for taking training courses is most often connected with their desire to diversify free time and master a new practice, not realizing the rational benefit for oneself.

As an additional hypothesis, we advanced the idea that the purposes composing elderly people's interaction with a computer and the Internet are rarely connected with lengthening activeness, for example, through inclusion in employment. Inclusion occurs, but predominantly within the leisure sphere.

The interview guide included three thematic blocks, helping the interviewer unveil processes of of elderly people's immediate interaction in the context of studying in class, as well as processes of independent computer mastery, daily user practices, and reasons for avoiding them.

The interview material was developed through the program Atlas.ti, intended for qualitative data analysis.

7.2 Familiarity with ICT: First Interaction, First User Practices, and First Fears and Mistakes Connected with Mastering Virtual Space

For older people, a computer does not usually appear in the family from their necessity or desire to use it. Most often, the computer is purchased for children, and then when they grow and begin life on their own the computer remains as a legacy to the older parents. Accordingly, the age of such machines might waver from 2 to 15 years, which can consequently play a significant positive or negative role in processes of mastering computer literacy.

Ending up one-on-one with a "not-so-new" but a still unfamiliar machine, older people finally begin to interact with it. As a rule, the main but unspoken motive for socializing over the virtual-computer space is due to children moving, the death of loved ones with which the person lived – in short, the feeling of loneliness that most elderly person encounter at some time.

At least in one interview, discussion circled around the fact that the respondent, lacking a home computer, decided to first take relevant classes, and to then decide whether or not to buy the technology. In all other occasions, the elderly respondents already had home computers, often possessed for years in disuse.

As already stated, most often older people's interaction with ICT begins with computers left behind by maturing children. As a result, initial ICT skills are developed with the help of relatives (children and other family members) and are formed as "the children showed." The computer might be literally inherited, for example, from husband to wife. In these cases, the computer is interpreted as property, belonging to a certain person. When this person dies, the computer becomes the property of the person who outlived them.

Senior citizens originally master many skills through the "trial and error" method, which does not qualitatively differ very much from the methods their children demonstrate. As a rule, speed in remembering keyboard configuration and the buttons that make computer work possible substantially differs between young and older people.

Even more rarely, senior citizens resort to reading self-help books, while the skills obtained amongst the elderly through work experience with CPUs are, as a rule, called too highly specialized or antiquated for casual use today.

As apparent in the quote, older people's familiarity with computers almost always happens through a mediator's help, functioning as the "installer" of needed programs and demonstrating simple, but hard to remember actions, and providing computer work. In the process of clarifying for what purpose the respondents used a computer before taking the classes, it appeared that a significant portion of elderly people did use a computer at work. However, computer literacy skills possessed at work and at home differ substantially.

Computer work in a professional capacity is strictly limited functionally to a selection of narrowly-specialized programs, needed to complete specific actions. In addition, having this set of skills, elderly people note the lack of "normal" daily user experience with simple programs and applications. And, in contrast, the skills that elderly people aim to develop do not relate to improving or developing already-possessed skills; these are two completely different knowledge types and ways of interacting with ICT.

The distinction between computer use practices "at work" and "at home" is connected in many ways with the presence and lack of control over these practices from the side. Limiting computer usage at work has always been connected with external objective circumstances: control over work time and its substantive content, the closed nature of the companies where respondents work(ed), narrow program selection, and lack of Internet access.

Another distinction – between real and virtual space – creates the absence of constant external control. Possessing computer work skills, older people seemingly open a new world unto themselves, attractive, yet full of dangers which they have difficulty protecting themselves from independently.

Thus, actively using television as a main source of objective information encourages elderly people to put forth the opinion that the computer is used as a secret observation instrument examining their life, in particular. It is noteworthy that the desire to protect oneself from the "all-seeing eye" makes elderly people act extremely resourceful, displaying initiative and interest (particularly significant in contrast with their desire to learn to use a computer):

The fear of hidden watching is an effect of the more rational fear of losing anonymity, which a small number of respondents indicated that they had encountered:

The objective risk that elderly people encounter when beginning to use a computer is the risk of downloading a virus. Resisting virus attacks often requires additional knowledge and installed programs, which not only elderly, but even advanced users don't always have. Apart from that, the fear of infecting the computer is connected with contacting a third party, i.e. the public acknowledgment of an error, causing an undesirable reaction for elderly people.

Efforts to gain computer literacy, on one's own or with minimal help of friend and family, are not always successful. Fears associated with using the new information-communication space, dissatisfaction with already developed skills, free time possession, access to courses at comprehensive social service centers and libraries… Usually, these reasons in particular form a need amongst elderly for enrolling in specialized courses. However, the goals respondents formed for such training may substantially differ.

7.3 Computer Literacy Courses: Goals and Motivation for Continued Learning

As mentioned earlier, older people often separate user experience gained for work-related purposes from the skills required for daily computer and Internet usage. Mastering the often difficult professional programs, study informants acted helpless in applying these skills toward simpler tasks. Thus, Internet search, online communication, and new diverse forms of digital recreation are becoming desirable but independently unattainable dreams for the elderly person.

As we stated at the very beginning of this section, loneliness and insufficient socialization both with loved ones as well as with the outside world are senior citizens' primary initiating factors for using the computer and Internet.

Study informants verbalized another basic goal for computer education as the abstract desire to "learn everything," coming from the childish formula for expressing curiosity in the face of newly-discovered world diversity: "I want to know everything!"

From our perspective, this very abstract goal amongst many study informants testifies to the fact that frequently there is no specific idea lying behind the desire to attend classes. Most often, elderly people are trying to diversify their leisure time, often without acknowledging that these obtained skills can be used for something beyond filling free time or as a new form of entertainment.

At a later stage, the study informants' array of goals does start to taper down (although not significantly) to the desire to "master Internet usage" and "master computer usage." Remarkably, older people, forming new goals, push away their previous computer interaction experience and the limits associated with it. For example, previously lacking Internet access at work motivates the desire to fill this void now by "mastering the Internet."

Earlier-obtained technical education and specific professional association allow some older informants to relate to the computer as a "machine" that they find interesting to assemble and disassemble. Alongside the more-or-less obvious "applied" skills that senior citizens obtain by attending classes, reflection on the topic of personal aging can also lead them. In this case, gaining skills for working on the computer and the Internet equate to methods of preserving personal activeness, in contrast to others lacking such skills, as a way to improve their image in loved ones' eyes.

Several older people who were questioned during our interviews had already reached the point in the basic computer literacy course where they had become "hooked" in the process and were preparing to "go to the next level":

Our expected motivation [for taking computer classes], linked to increasing professional skills by mastering computer literacy, turned out to be completely unpopular. Moreover, even the few informants who took classes with the desire to increase qualification complained about the lack of demand connected with old age. At the same time, even simply unqualified positions that do accept older people are complicated to access because taking classes does not guarantee confidence in one's abilities, acknowledging that the skills needed to work with a computer – for example, in a security position – are minimal:

Supposing that for the government dispersing free computer literacy courses amongst comprehensive public social learning centers and libraries was connected with popularizing electronic services, including amongst the elderly, then in this sense, based on our interview materials, they clearly have not effectively accomplished their goal.

Only one of our informants stated that she preferred going to common-application document centers to using online services. Obviously, even with state's great desire and opportunities to attract a maximal number of older users to electronic government sites, elderly people will continue to give preference to already-familiar methods of interacting with state institutions, exhibiting distrust toward new formats.

Thus, for example, paying apartment rent via impersonal contact with a computer is unlikely to replace a ritual monthly hike to the bank. For another thing, the availability of a bank card as a mediator in receiving electronic services also serves as a barrier for many senior citizens, creating additional distrust in this existing, accustomed, and understandable form of interaction (see Chap. 8).

After starting to attend classes and encountering certain difficulties, many older people still face the question of whether it is worth continuing education. Almost all continue their training, motivated primarily by the fact that "it turned out to be better and easier than expected." For some senior citizens, continuing computer training opens ever newer opportunities, compensating for earlier knowledge gaps.

7.4 Older People's Daily Interaction with the Computer: User Practices, Difficulties, and Reasons for Avoiding Several Actions

Elderly people's daily practices using a computer and the Internet little differ from any other user's practices. In this sense, the older person can hardly be described as a member of some sort of special user group. Like the majority of other people, seniors primarily master and use programs and applications designated for virtual socialization: e-mail, and the social networks VKontakte, Facebook, and Skype.

In addition, e-mail is most often used for formal socialization, for example, to place an order at a store, while social networks and Skype are used for contacting loved ones, relatives, and friends, and so forth. Moreover, social networks often become a means for older people to "revive" old acquaintances and a kind of "facilitator" for the time in life when they were younger and socialized face to face.

The most important result of continuing classes and simultaneously widespread use in practice is the opportunity for socializing and supporting connections to loved ones who might live in other countries or in a different part of the city.

Most often, the Internet serves older people as an informational-entertainment channel. Having mastered the classes, elderly people behave predictably: they read or watch the news, they follow weather reports, and they download or watch movies online – in a word, they use modern technologies as a new way of exploring free time. In this case, the computer becomes a more progressive, interactive alternative to television, and for some people even literally replaces the TV.

Logically resulting from the previous paragraph, older people use skills searching for information on a daily basis, including to seek thematic information that is specially addressed to them. For some older people, the Internet has become a convenient "housekeeping aid".

Medicalized leisure time is common amongst Russian retirees. If having free time before forced elderly people to endure huge lines for the doctor's office, now the negative costs of this pastime have substantially decreased. There is no need to even leave the house to take a referral to see the doctor or to get instructions for using some kind of medical device:

Referencing her groupmates' experience, one of the informants told about how many use the obtained skills for civil activeness, in particular, writing complaints. Undoubtedly, this result of seniors being introduced to computer literacy courses is productive and testifies to the empowerment opportunities by articulating a social position and including older people in an active civil life.

However, there were people among the elderly informants who use the computer and the Internet exclusively temporarily – in the context of learning in class, not rushing to apply the obtained knowledge to daily practice:

In the most general sense, the main struggle that older informants expressed is that "everything is new – this is always difficult." When speaking about the new, older people accentuate not just the modern technology's features and the necessity of mastering it "from zero," but also, more importantly, on their age and the confusing particularities of gathering information and learning. The difficulty of mastering the computer is connected to physical limitations and the general exhaustion typical of the body "wasting away".

From our perspective, a difficulty for elderly people, which was articulated to a lesser extent but is no less important, is the problem of concentrating attention on specific things and lacking the patience for assimilating new, frequently boring information. It is unlikely that this can be called a particularity of old age specifically, because children, for example, also often do not possess diligence and patience in absorbing more difficult information. Aside from this, many elderly people lack self-education skills or motivation for learning, but are prepared to receive "services"–they desire not to learn, but to be taught.

7.5 Elderly Interaction and ICT Through Computer Class Teachers' Eyes

In this segment lie the results of research dedicated to the characteristics of the content and effectiveness of computer literacy courses for an older audience from the teachers' point of view. We suggested that one of these methods of evaluating the effectiveness could be asking specialists to reflect on their own experience interacting with older people in the process of their training on the classes. From our point of view, these "experts" are capable of analyzing and reflecting on the content of the courses they teach, as well as receiving feedback from older people to determine the effectiveness of their own activity, at least in the context of the tasks posed to them.

The main research hypothesis was that computer literacy training as "the technology of socially including older people" is far from always leading to extended active work life, while it does encourage developing new social demands, especially in the leisure sphere. As an additional hypothesis, we proposed the idea that the very structure of the computer literacy courses offered by the state, as well as their content, does not stimulate older people's longer employment by mastering new electronic technologies.

The research task included evaluating the influence of new technologies of interaction between elderly people and society (in particular, computer literacy training) on their social inclusion. Under interaction technologies, we consider social services for the elderly that are legislatively and institutionally affixed and used by government social services working with older people.

We used expert interview as a method of gathering data, including three blocks of dynamic questions united into the following thematic blocks:

- The characteristics of offered services associated with developing interaction between older people and ICT
- The features including older people in an informational society's practices: barriers and opportunities
- Evaluating the effectiveness of offered services associated with developing interaction between older people and ICT.

Audiences of Courses Offered at Neighborhood Social Service Centers, and Older People's Motivation and Demands.

From the perspective of specialists working in social service centers' social and leisure departments, motivations amongst older clients for attending computer literacy courses are fairly diverse. Older people create different needs that, as they believe, they can satisfy by mastering new computer and electronic technologies. These needs are most obviously separated into the following categories:

- Need for socialization (communication with the help of new means of contact, including social networks);
- Need for increasing qualification (as a rule, a person forms this demand at a permanent work place);
- Need to follow fashion (desire to tell/brag about possessing progressive skills);
- Need to implement and confirm "care" from the state (it is important that the very fact of using free services, skills received as a result might not be used);
- Need (aim) to overcome loneliness (perhaps accomplished in the process of actually taking the classes – with the help of meeting other students, as after mastering existing skills in online space).

As "the utmost important need," computer class teachers named the desire to use "computer opportunities to the maximum," in the sense of how they themselves understand this. More likely, this understanding passes through individual experience, which ascribes several characteristics of the "average" user, possessing Internet search skills, downloading audio and video materials, payment services, ordering services and items, and so forth.

The need for increased qualification to use devices at work by non-working retirees was named as the most rarely encountered. At the same time, the most widespread need is the understandable desire to "be like everyone;" to not fall behind others. The desire to master elementary computer work skills can serve as a motivating circumstance, since "possessing a PC" is now a mandatory hiring prerequisite for any job. However, by themselves these skills do not encourage increased professionalism, but are rather akin to being able to "press a button" at a needed moment.

When necessary and with motivation and desire to spare, older people can continue training in classes, crossing over from the "basic" program to the "advanced." As a rule, this motivation arises in connection with the need to broaden skills encouraging increased diversification not of professional, but of older people's leisure practices.

Younger generations of relatives can be another important "driving force" in leading older people to take classes. As a rule, these previous generations of relatives, children or grandchildren in particular, become "guides" for older people in the "computerized world." The "younger" persons' motivations can also vary, and even the very speed of older people's "interaction" with the computer occurs both directly and latently" (Sergeeva and Paramonova 2012).

Relatives of elderly people can become interested in the opportunity to support long-distance communication; so that older relatives use the presents, which are now acceptable to give to people of all ages; in eliminating unused technology (through re-gifting or handing down older models "as an inheritance"), or in documenting family history through digitalizing old photographs (in which the role of "digitizer" is delegated to older people as the protectors of history, who are accommodating and gladly accept the monotonous and uninteresting occupation).

Additionally, some older people consider the chance to attend classes as an alternative to the daily life they currently possess, and which they do not find satisfactory. Some equate lacking skills for using modern computer technologies literally with the process of dying off or nearing death.

7.5.1 Classrooms Content, Opportunity, Demand, and Stimulating Obtained Knowledge and Skills Use in Labor Activity

The thematic structure of classes as a whole corresponds to the older people's (possibly, made up) demands. Accordingly, the tasks facing social services, libraries, and the state match the target group's desires. Nevertheless, the question arises: in which key does the government perceive support for the elderly? It seems that the government aims to nurture a modern user in the post-Soviet person, seeing it as the opportunity for elderly social activation. And these services are truly popular.

In this case, the state acts rationally, slipping members of any age or generation into a universal marketing strategy in which, in order to be like everyone, it is necessary to "consume", and in order to consume at the same level as everyone else you need to have the corresponding skillset (Ilin 2007). Constructing a new user group, almost as if "upgrading" the elderly, the state doesn't direct its efforts at the opportunity to use obtained skills in other areas of life, directing them instead to the consumer and leisure spheres. Older people actually gladly accept this ideology, agreeing that at this particular age they are "worth it,"[1] more so at the state's expense.

The beginning program is divided into two parts. The first part is acquaintance with the computer in general: with the operational system; with the different elements and subjects in a personal computer; and working with files and folders. By the way, one of the most complex topics is the file system and the folder hierarchy system, this is simple, we will hang on this capitally and return to it, because for some reason it is hard to perceive this non-representational existence of some virtual elements, as they shift amongst themselves and beyond.

In addition, teachers' duties do not extend to the probability of recent job-hunting or stimulating this desire, even if such a demand arises amongst students. On the other hand, in certain cases the teachers themselves acknowledge that the knowledge and skills obtained through the classes at social service centers are not oriented on professional activity use. Moreover, the number of hours devoted to practical lessons is insufficient for independent use later.

Thus, possessing computer literacy acquires unique features of "prestigious" usage amongst older people, creating a symbolic border between later the "illiterate" and "capable," but not on the level at which this skill remains a symbol, not acquiring a utilitarian weight/able to be utilized:

In essence, any of these circles possesses a teaching nature, but we are not an educational institution and do not hand out diplomas. If a person needs some kind of confirmation of passing the classes, then we give him a certificate on an officially stamped form with the director's signature and seal and indicating the number of hours and the name of the class passed.

Cooperation in job-hunting in our case consists only in giving a person service employment contacts or finding the phone number of the department of the organization in which our ward wishes to work.

7.5.2 Complications and Barriers Occurring in the Course of Training

As one of the main difficulties, teachers named the "imprint" left by the long period of Soviet power in which the students spent a large part of their lives. While many people attending classes lack skills working with technology, the main problem is that they experience psychological discomfort, fear in the face of the unfamiliar,

---

[1] "Because I'm worth it" – The most famous slogan of the L'Oreal Cosmetic Line ad campaign, laconically personifying in its time the representation of a person belonging to a consumer society about their well-being and its material fulfillment.

new, and incomprehensible, which was formed long before society's and the work-place's mass computerization.

In addition, even the teachers themselves can create "artificial barriers," contra-dicting the principals of budget education's openness for all older people and excluding individual groups of elders from the opportunity to receive access to elec-tronic education. In this context, a situation is constructed in which the teacher's personal choice and perspective form the social characteristics of the target group receiving services.

Opportunities also appear, because in continuing the beginning classes I have an Internet club where we are already plowing through web space based on these basic learnings. And there I have a lecture about electronic government, paying for Internet purchases, and communicating online. I also have a separate class that is designed for people who actually use the Internet.

The elderly people also try to use the opportunity to take the classes for free several times. This type of behavior fits fully into the strategy of the "insatiable consumer," who frequently does not consider why he or she needs repeated use of the same product/service. In this case, the behavior is determined by factors such as a service's constant representation in the market, its free cost and accessibility, as well as its growing allure by attracting an ever growing number of older people into informational space.

There is a possibility of situations when relatives interfere with older parents mastering the computer, distinguishing them "competitors" using excessive time in the precious place at the computer. Thus, older people mastering the computer can experience a negative and reverse effect in inclusion in informational space and the support network of social/familial connections.

Becoming active users, older people aim to decrease inequality, arising earlier due to their natural time limit using the computer. If older people could earlier share the computer with other relatives, now they need their own individual instrument for social inclusion, the right to which they are prepared to defend.

7.5.3   The Effectiveness of Classes and Its Criteria

The majority of teachers with whom we conducted interviews did not have a clear or moreover formalized representation of measuring the effectiveness of the ser-vices they offer. In most cases, feedback from former students occurs by collecting enthusiastic but not very specific reviews about taking classes, which are posted in groups on VKontakte created during the training process.

This speaks, first of all, about the fact that the service itself was not initially developed to have criteria for evaluating its effectiveness. In many ways, this situa-tion is typical for Russia in the sense that offering a service by itself, i.e. unlimited accessibility, frequently becomes synonymous with effectiveness, replacing poten-tial measurements of how necessary or important implementation is for its target audience (Grigoryeva 2011).

Quantitative indicators enter into the calculation: the number of Social Service Centers participating in the program of elderly computerization, the number of older persons attending classes, and so forth. Qualitative service evaluation is not even formed as a goal; the very fact that the state provides the opportunity for seniors to interact with modern technologies is important in itself.

After conducting a round of interviews in Saint Petersburg libraries, we noticed that libraries' very atmosphere, employees, and methods of organizing computer classes for older people substantially differ from those at Social Service Centers. The main difference was the teachers' relationship to their work, since implementing similar classes based in libraries is essentially still in the experimental phase. In particular, lacking Social Service Centers' leisure divisions, library workers do not have specialized experience communicating with older people or organizing corresponding events and activities.

Obviously, these conditions, complicate the process on one hand, while on the other hand introduce unique features. Thus, teachers in libraries often approach choosing training techniques more creatively and reflect more with regard to their activity and how the classes can encourage older people's social inclusion.

To a certain extent, the actual training program also differs. Several teachers even noted that older people come to them, already having passed classes at Social Service Centers, but still wanting to learn again "from zero."

In spite of the less explicit accent on mastering Internet technologies in favor of more careful study programs for creating text documents and electronic presentations, teacher-librarians enumerate an even wider spectrum of skills that older people master in their classes.

In several cases, the skills gained give entirely specific results that not only encourage including elderly people in virtual social networks, but also influence their "real" life. Moreover, the composition of older people and their social characteristics likely also differ from those of people who take classes at the Social Service Centers.

Expanding the spectrum of leisure practices and practices using skills mastering the computer to improve older people's personal life is little connected with continuing labor motivations or elders' additional professionalization as the result of taking classes. Furthermore, lacking an original "order" for serving older people, librarians evaluate the potential for using obtained skills for increasing qualification in the more "everyday" sense.

Like in the Social Service Centers, no one measures library courses' efficiency, it lacks criteria, and feedback is received exclusively in the form of thank yous and congratulations through virtual social networks. In essence, there are no criteria given from the "top" relating to the effectiveness of this type of service.

Well, they, maybe, are asscribed in some normative documents in relation to this question, but to say specifically from who (I am speaking even not just for us, but for the city libraries as well) there were no criteria to evaluate the effectiveness of training, in principle. For us, the main criterion is our audience. Naturally, we try to identify effectiveness because each teacher is required to evaluate lessons. We save all these documents, we record everything because, of course, people's opinions are

Table 7.1 For what reason do you access the Internet? (as a % of the number of respondents)

Purpose of use	2009	2014
Information search	19	84.4
Reading news	5	68.8
Corresponding with friends via e-mail	7	18.8
Skyping with close friends and family in other cities/ countries	2	21.9
Interacting on social networks: Vkontakte, Odnoklassniki, Facebook, Myspace, Myworld, etc.	2	22.3
Reading books and periodicals	2	–
Online shopping	–	15.6
Making appointments with a doctor and other medical services		–
Searching for and ordering inexpensive tours or tickets	2	–
Telephone or utilities payment	2	–
Money transfer	–	–
Watching movies	–	15.6

important. In addition, there have been times when they told us that they believed that something else needed to be included in the basic program.

In some cases, the very instance of attending classes, and not just the skills received at them, becomes "leisure in itself," a new method of spending free time for older people (Abankina 2005).

Our other studies dedicated to older people's ICT practices also show that seniors are interested in a fairly narrow skill spectrum. For example, using e-mail as well as Skype and social networks quickly increased. Accordingly, surveying approximately 300 older people in Saint Petersburg in 2009 and again in 2014 gave us the following results (Grigoryeva and Chernyshova 2009; Grigoryeva and others, 2014a, b) (Table 7.1).

We see that the convenience of e-mail and Skype is highly valued by older people, and these services' user numbers have significantly grown over the past 5 years.

7.6   Conclusion

Our research confirmed the proposal that the leading computer training programs for older people in Saint Petersburg social service centers and urban libraries were, on one hand, not originally designed to improve their opportunities in the labor market, although this task is included in the concept of continuing education. Furthermore, teachers do not pose such tasks to themselves or their class-takers.

On the other hand, the elderly people going to classes are rarely motivated by professionalization, increasing qualification, or continuing labor activity. However, using the computer becomes a widespread consumer practice, confirming an older person's the "modern-ness" and training in such skills has become trendy amongst the elderly.

The computer literacy class program offered in libraries has some outstanding features. The very atmosphere of the library – initially an enlightening, intellectually-leisurely institution – creates conditions for even more class "rollout;" not just as an instrument for comprehending new practices, but also as a full-blown leisure form, for the inclusion toward which older people can persistently strive.

Thus, Social Service Centers's example, as well as libraries', demonstrates that the originally-educational initiative, lacking clear criteria for evaluating effectiveness, is transformed from supporting the opportunities of elderly employment into an instrument for expanding their social connections and communication.

Data from studying older people shows that the range of motives for immersion in educational practices is generally extremely wide: from narrow utilitarian goals to philosophical-world views. For the state, which finances the training, and the teachers, information about retirees' real motives has a wider significance, as certain discrepancy among proposed results is observed to be real. But even the narrow tasks of training, ultimately, also serve as a form of social inclusion for the older generation.

References

Abankina, T. 2005. Ekonomika zhelanij v sovremennoj "tsivilizatsii dosuga". *Otechestvennye zapiski* 4 (25).

Bauman, Z. 2004. *Wasted lives. Modernity and its outcasts*. Cambridge: Polity.

Bourdieu, P. 2005. *Sociologiya socialnogo prostranstva* (Trans: Shmatko, N.A.). Moscow: Institut ehksperimental'noj sociologii; St. Petersburg: Aletejya.

Butueva, Z.A. 2014. Osobennosti social'nogo samochuvstviya lyudej starshego vozrasta, nahodyashchihsya v trudnoj zhiznennoj situacii. *Vestnik Buryatskogo gosudarstvennogo universiteta* 5: 127–129.

Castells, M. 2009. *Communication power*. Oxford: Oxford University Press.

Elyutina, M.E., and E.H. Chekanova. 2003. Pozhiloj chelovek v obrazovatel'nom prostranstve sovremennogo obshchestva. *Sotsiologicheskie Issledovaniya* 7: 56–73.

Giddens, A. 1990. *The consequences of modernity*. Cambridge: Cambridge University Press.

Grigoryeva, I.A. 2011. Razvitie socialnoj raboty v rossijskom obshchestve potrebleniya. *Zhurnal sociologii social'noj antropologii* 5: 287–297.

Grigoryeva, I.A., and S.P. Chernyshova. 2009. Novye podhody k profilaktike socialnogo isklyucheniya pozhilyh. *Zhurnal sociologii i social'noj antropologii* 2: 186–196.

Grigoryeva, I.A., L.A. Bershadskaya, and A.V. Dmitrieva. 2014a. Na puti k normativnoj modeli otnoshenij obshchestva s pozhilymi lyudmi. *Zhurnal sociologii i social'noj antropologii* 3: 151–167.

Grigoryeva, I.A., A.S. Bikkulov, and G.M. Cinchenko. 2014b. Starenie, mezhpokolennye vzaimodejstviya i zanyatost lyudej pozhilogo vozrasta. *Upravlencheskoe konsultirovanie* 12 (72): 101–110.

Ilin, V.I. 2007. Potreblenie kak diskurs. *Zhurnal sociologii i socialnoj antropologii* 10 (1): 3–26.

Kononygina, T. 2006. *Geragogika*. Orel. http://lit.lib.ru/t/trushnikow_d_j/text_0250.shtml. Accessed 6 Dec 2017.

Kovaleva, A.A. 2008. Samosohranitelnoe povedenie v sisteme faktorov, okazyvayushchih vliyanie na sostoyanie zdorovya. *Zhurnal sociologii i socialnoj antropologii* 11 (2): 179–191.

Krasnova, O.V. 1998. Stereotipy pozhilyh i otnoshenie k nim. *Psihologiya zrelosti i stareniya: vesna*: 16–28.

Okinawa charter on global information society. 2000. http://www.g8.utoronto.ca/summit/2000okinawa/gis.html. Accessed 6 Dec 2017.

Pisarev, A.V. 2004. Obraz pozhilyh v sovremennoj Rossii. *Sociologicheskie issledovaniya* 4: 1–20.

Programma razvitiya OON. 1997. Ezhegodnyj doklad http://www.unece.org/fileadmin/DAM/publications/AnnualReports/1997-1998_Interactive_Annual_Report_RUS.pdf. Accessed 6 Dec 2017.

Riva, Giuseppe, Paolo Ajmone Marsan, and Claudio Grassi, eds. 2014. *Active ageing and healthy living: a human centered approach in research and innovation as a source of quality of life. In Studies in health technology and informatics.* Amsterdam: IOS Press. https://doi.org/10.3233/978-1-61499-425-1-1 978-1-61499-425-1.

Saponov, D.I., and A.A. Smolkin. 2012. Socialnaya ehksklyuziya pozhilyh: k razrabotke modeli izmereniya. *Monitoring obshchestvennogo mneniya* 9 (10): 83–94.

Sergeeva, O., and V. Paramonova. 2012. Praktiki kak sredstvo razlicheniya: issledovaniye kompyuternogo opyta pozhilyh lyudey. *Izvestiya VolgTGU*. 11 (8): 67–71.

Smirnova, T.V. 2008. Pozhilye lyudi: Stereotipnyj vozrast i socialnaya distanciya. *Sociologicheskie issledovaniya* 8: 80–87.

Vershinskaya, O. 2015. *Cifrovoj raskol – novyj vid ehkonomicheskogo neravenstva?* http://viperson.ru/wind.php?ID=637647. Accessed 6 Dec 2017.

Chapter 8
Internet Space as a Platform for Studying Elderly Social Inclusion Opportunities

The Internet's ever more frequent infiltration in the social activity sphere gradually makes a means for increasing different social groups' social inclusion. It is believed that the Internet's especially great potential exists in solving the task of raising older people's social activeness. Special sites exist for seniors' communication, computer literacy training, and informing older people about electronic services, etc.

Internatization trend going together with the population aging raises many opportunities for enhancing the elderly social inclusion in different fields.

Currently, there are a multitude of research work dedicated to studying Internet and social network audiences, yet the category of third age people frequently remains outside the focus of these studies.

Of course, age particularities play an important role in using information technologies, the Internet in particular (Grigoryeva et al. 2014a). Despite the multitude of risks: medicinal (Kroshilin and Medvedeva 2011), in the sphere of data security (Yashina 2014), and others, ICT makes life easier and helps increase the quality of life. Skills of using new technologies lets older people open themselves to a multitude of new opportunities and even change their relationship to several things in life (Obi et al. 2013). Despite the fact that retirees yet still are not the most active Internet users, separate studies proclaimed that older people begin to more actively use new technologies with faster Internet, having the opportunity to access the Internet at any time, and if relatives and acquaintances help them to master new things (Laconi et al. 2015).

A number of scholars have already noted the positive effect of ICT's influence on elderly people. For example, American scientists evaluated the interconnection between older people using ICT and their depression level by using methodology combining regressive analysis and preference evaluation. Researchers proved that using Information Technologies lowers elderly people's level of depression by 20–28% (Vacek and Rybenska 2015). Other researchers confirm: using new technologies enables social interaction amongst retirees, an increased life satisfaction level, and helps them to deflect thoughts connected to health problems (Choudrie and Vyas 2014).

© Springer Nature Switzerland AG 2019
I. Grigoryeva et al., *Elderly Population in Modern Russia*,
https://doi.org/10.1007/978-3-319-96619-9_8

At the same time, the phenomenon of third age people using electronic state technologies and electronic participation instruments is still little studied. Currently, there have already been attempts to measure older people's relationship to electronic services. According to results of a study that was conducted in Finland, age does not impact electronic government service usage level (Cotton et al. 2012). In Germany, according to Vinkhoven's data, advanced aged people reluctantly participate in open state projects and public crowdsourcing (Challenge.gov, Maerker Brandenburg).

Japanese scholars distinguished factors influencing older people's participation in electronic government projects and electronic participation. These factors are: volunteers advancing these projects, broadening older people's social network usage, and creating volunteer communities to help older people master new technologies (Troisi 2007).

According to data from studies conducted daily by the company TNS Web-Index, in 2015 more than half (52%) of men from the ages of 55–64 years use the Internet, and this indicator is almost two times lower (27%) in the "65 and older" group (TNS Web Index, 2015). In addition, almost half of the women (49%) from the ages 55–64 years use the Internet, while only 17% do so in the age group "65 and older." Meanwhile, the Internet audience amongst older people in large cities (Moscow, Saint Petersburg) reaches 60–63% for people ages 55–64 and 28–50% for persons over the age of 65. These studies also recorded mobile Internet use over 18% of the Russian population ages 45 and older, while in Moscow this figure reached 29% and in 23% in Saint Petersburg over the given group.

The given data testifies to the Internet's incorporation in older people's practices, as well as to the appearance of opportunities for that population group to use information technologies for entirely diverse objectives: communication, transaction, receiving services, electronic participation, making administrative decisions, and voting on portals for citizens' electronic free will expression.

Studying older people's social activeness online takes on particular relevancy in this context, as well as the various mechanisms for activating seniors' social behavior based on using new technologies. The research interest was focused on the following interconnected practices: (1) using e-resources for education; (2) broadening employment opportunities based on using IT; (3) increasing political and civil activity through distributing state e-services; (4) discussing important issues at forums.

8.1 Educational Resources for Older People: Evaluating Demand

According to the data from international studies, IT devices allow people of an older age to further feel themselves active and even younger than their years. As noted above, the life expectancy at retirement has noticeably increased in all developed countries, which means the health improvement. Another reason for expanding life space, according to research by the International Institute of Applied Systemic Studies (Ryabova 2015), is the usage of modern devices. For this reason, the Western retirees today have become "younger" in the mental plane by 4–8 years in comparison with people of the same physical age have been 8 years ago. Scholars led cognitive tests over 2000 older people in England and Germany in 2006 and 2012, and came to the conclusion that the average IQ is increasing both in selected countries, as well as the world as a whole. The simplified access to more qualitative education resources is one of the main explanations, as well as improved nutrition, and healthcare for all the wider layers of the population. However, the research results showed the correlation between active modern technology usage and preserving and even improving mental capabilities.

In Russia, elderly people's internet usage is given special attention (Lysenko and Fedoseeva 2014; Panina and Pavelieva 2014). It is possible to observe the existence of educating older generations to work with new technologies (computer literacy classes, concessional centers, all-Russian and regional competitions and championships) institutionalized practices, as well as developing online platforms for retirees based on citizens' initiatives (Timurovtsy Informational society, Baba-Deda, All Years, and others). A special research interest represents the adaptation level of new resources to the elderly needs, and the demand level among the target audience.

According to TNS Web Index (November 2017), older users (55+) accumulate 14.2% among Russian Internet-audience. For comparison: in February of 2014 this percentage was 8.2%, and in 2013 it was 6.6%. From the statistic data, there are more and more older users among the Internet audience, and in the future older people in particular will become one of the main sources for growth. In addition, according to data from VCIOM, the Internet is used by 32% of Russians within the ages of 56–60 years, 13% within the ages of 61–72 years, and 6% of ages of 72 years and over, while 7% of the 56–60-year-olds surveyed have used the Internet for a long time.

The analysis of specialized Internet resources demonstrated their diversity. It is possible to see socio-informational (Baba-Deda, Pensioner.info), consultative (Golden Age), educational (PC-Pensionera, Third Age Univesity, After 50), and other spheres. We gathered the most active resources. Specialized Web Resources for Older People in Russia are:

- Third Age University (http://u3a.ifmo.ru/);
- Socio-Consultative Resource for Older People of All Years (http://vsegoda.ru/);
- Petersburg Pensioner Informational-User E-Magazine (pensionerka.spb.ru);
- Modern e-Jurnal for People 50+ with an Active Lifestyle (http://pokolenie-x. com/?p=3150);
- Independent Project "Golden Age" (http://zolotoy-vozrast.ru);
- Virtual School – Pensioner of the Future and Present (http://www.ypensioner.ru).

However, the research didn't notice the significant level of their convenience for the target audience. The study was concentrated on the Thirf Age University (developed at ITMO University, Russia in 2012).

This portal represents a qualitative and accessible distat learning. The portal combines training courses with multimedia content in the following areas: financial security, creative photography, poems and prose, stage drama, art, e-government and e-services, ecology, religion, Internet, volunteering etc.

Accroding to Internet-statistics, this resource has gained increasing popularity over the years. In particular, in 2015 the number of users increased by almost 10 times. An average number of portal users reached 320 people a day. An average rate of "Pensionerka" (http://www.pensionerka.ru/), which is the personal site of one pensioner and does not contain any kind of institutionalization attributes (educational courses, regulatory clearance, and so forth), is 168 visits a day, and every visitor opens 4–5 pages on the site. Web-site "New Pensioner" (http://www.pencioner. ru), created with the support of the Federal Agency for Print and Mass Communication, calculates 1200–1800 visits a day. This data confirms that PR-support (the site is presented also on all social networks) plays a significant role in supporting the level of resource use. Also, the search engines Google and Yandex are the main sources of the visitors. Analyzing search queries showed that the target audience's number is quite law (less than 1% of queries belonged to the topic).

Data from the service Yandex Metrics proved the fact that the portal's target audience did not generally reach the site. Amongst portal visitors, users 45 years + did not exceed 12%. Women predominated amongst portal users (69%), but these are mainly representatives of younger age groups.

8.2 Search Inquiries by which Users Come Across the Third Age University's Portal

In 2015 we conducted an online survey to find out the barriers that the elderly faced with on the specialized resurces. The survey was conducted among the Third Age University users aged 55 and over. The questionnaire was published at the portal, and 105 people took part in this study. In the research we studied the intensity of Internet site usage (level of online engagement); the most popular site topics amongst older users; the most important problems that older users encounter; the frequency of using state institutions' sites as well as state and municipal service portals. During the study period, invitations to the survey were shown only to users from Saint Petersburg, for which an addendum was written to determine the participant's city. IPgeobase (http://ipgeobase.ru) was used to determine city location.

The study showed that more than half of surveyed older representatives (58.6%) spend up to 3 h a day on the site, while 14.7% spend from 6 to 9 h. According to the survey results, the main reason for using Internet sites is searching for information online (90%). Also the relevant motivates include reading news (67%), communication via social networks (60%), job search (30%), receiving government services (28%), communicating via forums (16%), and ordering medicine (11%). As additional answers, respondents also chose ordering train tickets, as well as various events over the Internet.

It is worth noting that technical compatibility (cross-browser compatibility) and site speed play an important role when working with a website. The software tool Google Page Speed Insights measures webpage loading speeds. The URL is checked in two ways: with the help of a desktop and mobile user agent. The PageSpeed evaluation can amount between 0 and 100 points (where 100 is the maximum value). Based on the evaluation results, specialized web resources for older people were discovered to be far from all possessing a comfortable speed for page-viewing. According to data from other studies, slow page loading is a reason why a user becomes frustrated, leaves a resource, and doesn't visit it again.

The respondents were asked to show what they would like to change on visited Internet resources. Based on this study's findings, ads on sites are the key frustrating factor for older people. In addition, the majority of survey respondents complained about complicated registration procedures on sites and portals (registration by email, registration to receive state services, etc.).

The study brought the following conclusions of web-site usage:

1. The Internet is becoming more popular among older people, the ways to use it become more and more diverse. Internet appears as a daily practice of communication.
2. Informational and educational web-sites predominate amongst the multiple resources for older people, created both from the side of authorities as well as from civil initiatives. However, a small amount of resources is actually convenient for the elderly. The greatest barrier for further use of a resource is connected with annoying ads, as well as insufficient reference information for working with a portal.
3. According to the results of several studies, Internat usage by older people allows them to increase brain activity, meanwhile lowering the risk of cardiovascular diseases (Internet polezen..., 2014). While maintaining positive dynamics of using different Internet services, we consider a positive influence on older people's health level that allows them to remain socially active for a longer time.
4. Comparative sample analysis of visits to web-resources made a conclsion that institutionalized resources were largely popular, with three times the number of original visitors than other enterprising informational or educational sites.

8.3   E–Government Technologies and E–Participation Tools: New Opportunities for Older People

With the aim of identifying the demand for electronic government technologies over the population, a sample survey was conducted in Saint Petersburg in autumn 2015. The research assumption concerned the increase of political and civil activity through distributing state e-services.A non-repeating stratified sample was used, with the reliability level at 95.4%. Data for pre-retirement age groups (46–59 years) and early retirement age groups (60 years and older) were used to prepare this material. In total, 125 people were surveyed in both categories. The survey was conducted in two ways:

- Face-to-face survey in places of state services provision (multifunctional centers, Pension Fund branches in different city neighborhoods),
- Online questionnaire placed on Internet-resources.

The majority of respondents participated in the face-to-face survey. According to the survey results, almost half of the respondents seek state services no less than twice a year. In addition, 72% of those surveyed individually go to government agencies to receive required services, 63% also visit multifunctional centers offering services, and approximately 8% seek needed information in service call centers. Electronic means of receiving state services are less popular at the given moment amongst St. Petersburg inhabitants ages 46 and older: approximately 14% used a Unified state and municipal service portal, and only 5% used a Portal of St. Petersburg state and municipal services.

Amongst the surveyed people who were informed of state service portals, almost 10% found them via Internet search, another 11% had friends and relatives tell them about the portal, almost 7% saw an ad for the portal in subway or multifunctional centers, and only 3% read announcements in places services' provision.

Of the few people who had already used the portal, the majority (7%) sought information about services they were interested in on the portal, contact information for specific state organizations (4%), submitted e-applications for public services (5%), downloaded application forms for supplying public services (3%), sought legislative acts (4%), or contacted IT service. In addition, 2/3 of people using the portal were satisfied with its work and only 1/3 said that they did not find the portal's work fully accommodating.

In addition to e-services' demand analysis, we looked into e-participation portals for political and civil activity of the elderly. We selected "Russian Public Initiative" (https://www.roi.ru, created in 2013) that placed civil initiatives of the Russian Federation on federal, regional, and municipal levels.With the help of the portal, citizens could direct authorities' consideration to initiatives on federal, regional, and municipal levels (Bershadskaya et al. 2013a). In order to vote on the Russian Social Initiative portal, a person must register on the Unified Portal of the Russian Federation's State and Municipal Services, which involves pre-registration and authorization on the website (www.gosuslugi.ru).

The mechanism for working with initiatives is the following: the petition must collect no less than 100 thousand votes before it will be considered by a government task force consisting of 34 people. This expert group is composed of executive and legislative power representatives, business communities, non-commercial organizations, scientific organizations, and various funds. This portal maintains a high anonymity level: the names of the citizens who initiated the petition as well as those who voted for it are inaccessible to other users. At the given moment, more than 9 thousand initiatives are up for voting.

At ITMO University's department of state information systems management, an automated system was developed for analyzing e-petition portals (http://analytics. egov.ifmo.ru). The system makes it possible to trace new petitions as they appear in real-time based on various topics, determine their voting course, as well as build a prognosis in relation to whether the petition will overcome the necessary threshold of votes to move on to the expert group's consideration.

Using this system, we distinguished initiatives on topics associated with retirees. To accomplish this, we assembled a set of key words and using them implemented an automatic search over the database of submitted petitions. During the study, we concluded that over the portal's existence 75 initiatives were published on the target subject (many of which were dedicated to the topics of retirement and pension reform, etc.). We determined that, on average, 2–3 new initiatives have appeared monthly over the last 2 years in relation to the given subject matter.

The most popular petitions (those gathering the highest number of votes) were the following:

– Not to pay out pensions to retirees from power-wielding agencies until they reach the retirement age designated for citizens by federal legislature (55F16712), submitted 01/12/2015, votes "for" – 1047, "against" – 569;
– To create an expert state organization for checking the accuracy of amounts credited for pensions and/or financial aid (66F17045), submitted 01/13/2015, votes "for" – 730, "against" – 125;
– To annually index citizen's pension contributions by the inflation level plus 4% (63F17349), submitted on 01/18/2015, votes "for" – 1396, "against" – 121;
– To index the salary for workers of all institutions independent of the forms of their personal property and pension payments no less frequently than once a quarter on the actual inflation growth percent (62F17501), submitted by 01/22/2015, votes "for" – 1732, "against" – 178;
– To produce an annual pension reassessment based on the total payments flowing from working retirees' payroll to the RF Pension fund (77F17335), submitted 01/27/2015, votes "for" – 762, "against" – 91;
– To establish free passage for RF retirees on public social transport across all Russian territories (52F18204), submitted 02/08/2015, votes "for" – 1160, "against" – 192.

In addition, it is imperative to emphasize that due to the petition authors' and voters' anonymity, there is no opportunity to clarify whether older people submitted the petitions and voted. The study showed that none of the studied area's initiatives gained the necessary quantity of votes to transition to the expert group voting. The dynamics of the collected vote sample, highlighted a trend characteristic for initiatives on all topics: most votes for the initiative are collected within the first 2 days after being published, before there is a sharp decrease in voting activity.

Examples of successful initiatives, reaching results via changes in legislature, show that civil activists extensively use PR-companies to support their initiatives, thereby fueling interest over a longer period. These strategies can also be suggested to advance initiatives on topics linked to older people.

State services is an important issue for older aged people: up to 93% of people in that demographic appeal for services several times a year. Furthermore, only 14% used the federal e-services portal did this, and 5% total used the Saint Petersburg state services portal. Considering that older people frequently receive information about these portals from acquaintances and relatives, as well as through Internet search, one can talk about demand emerging amongst them to use online state services.

The study revealed that at the research period older people in Russia do not actively use online instruments and participation platforms. On the petition portal, the portion of initiatives on topics affecting older people's interests constitute less than 2%. Moreover, most petitions observed are aimed at changing pension legislature.

Meanwhile, older people make up 37% of Russia's active electorate and attend elections more regularly than members of any other age group. The Russian Party of Pensioners was founded in 2012 and has its own individualized program, objectives, and active participants, even though it is not represented in the State Duma. Alongside the party, there are also noncommercial active retiree organizations, for example, the "Silver Volunteers" organization which has attracted more than 400 active volunteers ages 50 years and older. These facts testify to older persons' enormous potential for participating in social life.

It is imperative that older-aged people's opinions be heard when making political decisions, especially those that concern them. In this regard, government bodies on federal, regional, and local levels should particularly pay attention to specialized programs to popularize e-government and online participation instruments amongst older people, as well as to support the elderly in using new technologies (Grigoryeva, Dmitrieva, Grigoryeva et al. 2014b).

8.4 Discussing Pension System Reform Questions in the Russian Federation Through the Mirror of Social Media

It is difficult to evaluate how the pension reform system in Russian influences older people's social exclusion and the extent to which the pension system is adapted to changing economic and demographic realities. Traditionally, Russian citizens, both older and younger, perceived the pension system as a fundamental and unquestionable social guarantee, although experience over the last 25 years has clearly demonstrated that while the system has been remained intact, the compensation for the lost income guarantee has become fuzzier.

Pension reform has existed in in Russia and been discussed amongst both specialists and the general population since 1989. Discussion about proposed innovations, either stimulating or quieting, generally shows that the population poorly understands the purpose of the new changes and does not plan to alter savings behavior. Specialists presume that the "aging threat" emerges in the context of "the unchanging nature of savings behavior over the course of a life cycle" (V poiskah..., 2015), when people persistently extrapolate "Soviet" behavior models on a new socio-economic reality, although the irrationality behind this is already perfectly obvious. Yet, expectations in relation to state support still do not decline.

The Pension Reform of 2001 concluded with the personified accounting system's decline, blurred pension indexation rules, and the increasing randomness of current indexation solutions, as the reform architect, renowned economist M. E. Dmitriev, stated long prior (Dmitriev 2005). In a country where there is a significant "grey" employment sector, accounting for pensions in accordance with transferred employer contributions would imply leaving a significant portion of people without payoff upon reaching retirement age.

In 2008, a path was chosen to defend their interests, so pensions for people with five years' seniority differs little from pensions of those with over 20 years' seniority. The state protected the rights of those in the "grey employment" zone and discriminated against conscientious tax-payers. This, in our opinion, became yet another source of trust violation between the government, employers, and employees, because these people themselves draw closer to the point of leaving the working community for the retiree community.

As it appears, the state reformulated the pension system task from "receiving a sufficient pension" to "least time and labor contribution to pension." A portion of the population implements this strategy by seeking work that offers different service lengths with the right to early retirement, or seeking opportunities to register for disability, which also provides the legal possibility not to work and offers a small monetary payoff. Another portion of the elderly perceive retirement as social status loss, although many say that they continue working because "one cannot survive off a pension."

And at present, continuing pension reform is a constant discussion and argument subject. Based on data from a "Public Opinion" Fund survey, in 2013 56% of Russians stated that they were informed to certain extent about pension reform (Opros FOM, Obi et al. 2013). Furthermore, many respondents had difficulty evaluating pension innovations because they did not understand their essence, but negative reviews predominated among those who did offer an assessment. In addition, almost 70% of Russians believed that a pension in specific was becoming the main resource capable of provisioning old age (Opros FOM 2012).

A contradictory situation has thus developed: earlier surveys showed that 80% of the population did not want to increase the retirement age because pension is perceived as a fully-deserved additional monetary supplement upon reaching 55–60 years of age. But now the attitude is changing: "Many working-age Russians do not support the proposal to increase working seniority to 35 years to receive a full pension. While the majority approves the idea to increase the threshold for the right to a labor pension from 5 to 15 years" (Opros FOM 2012).

Per VCIOM data (Nevinnaya 2015), 80% of Russians do not support such innovations as increasing pension age, 75% are against canceling pension payout to working retirees, and approximately 60% oppose "freezing" retirement savings. We see that the data from VCIOM and FOM about increasing pension age correspond, and the minimum service length required to receive a work pension (5 years since 2008) seems, truly, to already be too small. This can attest to the fact that people have become interested in acknowledging the importance of labor participation and preserving the "labor," earned pension.

Many people also do not want to retire within the first five years, i.e. before 60–65 years (Maleva and Sinyavskaya 2008). "If in the past most people aimed to stop working, now there are increasing numbers of those who do not want to retire, as well as those, for whom the question of retirement is not even plausible. The first group includes civil servants, judges, professors, and scholars, who are constantly righting for the right to work beyond the established limit. Under their pressure, relevant changes are periodically introduced into legislature. The number of people are growing who work in liberal professions for as long as they can allow themselves (Maleva and Mau 2013).

Social media is a significant resource for citizens' discussion, and the pension topic is also covered in this online space (Bershadskaya et al. 2013b). We studied the pension reform discussion in Russian social media with the use of IQ Buzz tool, covering the period January 2010–May 2015. The research found 75,383 messages of the following types: comments, forum messages, microblogs, videos, notes, news, posts.

Over this period, we observed growing activeness in discussions on the given topics and several informational surges, associated with the related newsbreaks. Traditionally, pension payments increase on February 1 and April 1, and discussion surges associated with these turning points are observed yearly. In 2013, a Federal Law was adopted on December 28 No. 400-F3 "On Insurance Pensions," while discussion on this law peaked over December 2013 and January 2014.

Since January 1, 2015, in accordance with the Federal Law enacted December 28, 2013 No. 400-F3 "On Insurance Pensions" and the Federal Law enacted December 28, 2013 No. 424-F3 "On Cumulative Pensions," labor pension based on the age is assigned by a new pension formula. The related discussion has been observed since January 2015.

Also, over the entire period studied, we noted active discussions associated with the working retiree pensions. In particular, the first questions were actively discussed in July 2013. Moreover, question discussion continued in 2015 when the government proposed curtailing the pensions of working retirees whose salaries exceed one million rubles a year. Based on the data from a nationwide Russian sociological survey, the citizens are divided in relation to the given government plan (46% support the initiative, 44% are against it), but people more often support initiatives than express disapproval (FOM Survey 2015).

This support is a very strange, but serious testimony to the population's liberalized views, i.e. the transition from assurance that a pension is earned and cannot become alienated, to the conviction that, first and foremost, the poor and non-working retirees can count on a pension. Thus, it is impossible to rely upon population survey data while trying to find a rational reform option, considering opinions change quickly and depend on a number momentary circumstances.

Messages on the topic "Retirement" are published predominantly on the social network VKontakte, as well as in blogs and microblog services. The most popular resources for the discussion were the following: VKontakte, LiveJournal, Twitter, LiveInternet, Vladmama Forum, Google +, Banki. Ru, Mail.ru.

Gender distribution shows that men (56.7% of authors, composing 58% of messages) speak out more actively about the topic than women (43.3% of authors, composing 42% of messages) (gender determination – 60% of all sample).

Since the beginning of 2011, the discussion audience, consisting of people linked to retirement as well as website users who viewed these discussions, constituted more than 6.8 million people. Citizens of ages 26–35 displayed the largest interest in the studied topic. The topics covered receive extremely low responses from people over the age of 56, considering low overall low activeness amongst people of this age on websites and social networks. In addition, there are essentially no responses from children (those younger than 16 years).

Studying social media revealed that more than half of all published texts on the topic at hand are introduced by authors of ages 26–45 years.

It is worth noting the messages' and documents' geographical distribution by Russian regions. Moscow (16,476 messages) and Saint Petersburg (7257 messages) occupy first and second place, accordingly. The following offer 1000 to 2000 messages per region: Primorsky and Krasnodar Krai, Moscow, Sverdlovsk, Nizhny Novgorod, Samara, Novosibirsk, Rostov, Chelyabinsk and Volgograd oblast, the Republic of Tatarstan and Bashkortostan. In other regions, the examined topic was discussed less actively.

***

Analyzing a wide spectrum of newsbreaks allowed us to draw the following conclusion: the public actively discusses official news, events, and dates connected with pension reform. One can thus surmise that the population is becoming actively interested in pension rights formation and continues discussions on the present with ever growing interest.

Population survey results, conducted by large sociological centers, showed the ambiguity of evaluations on current reforms amongst different social groups. Authors believe that studying discussions slices on social media can operatively reflect the variation in users' assessments and serve as a basis for informing the population more thoroughly about new steps in each reform, since social media acts as a mirror, reflecting public attitudes.

In addition, pension reform discussion analysis disclosed that the public is worried about not just by the fate of pensions and pension accumulations, but also by the overall economic situation in the country. In particular, 43% of non-working retirees would continue their work experience after reaching pension age, but do not have the opportunity (Opros FOM, 2014). This data emphasizes the employment issue for people retiring who do not wish to lose their activeness.

Subsequently, the situation associated with evaluating elderly retirement prospects and employment in modern Russia (in the context of how continuing working influences the social exclusion) is contradictory and requires further study.

References

Bershadskaya, L., Chugunov, A and Trutnev, D. 2013a. E-participation development: a comparative study of the Russian, USA and UK e-petition initiatives. In *7th International Conference on Theory and Practice of Electronic Governance (ICEGOV 2013) Proceedings*, 73–76.

———. 2013b. Social media research: Network users' activities as a reflection of real life. In *CeDEM13. Proceedings of the International Conference for E-Democracy and Open Government*. 22–24 May 2013, 395–402. Austria: Danube University Krems.

Choudrie, J., and A. Vyas. 2014. Silver surfers adopting and using Facebook? A quantitative study of Hertfordshire, UK applied to organizational and social change. *Technological Forecasting and Social Change* 89: 293–305.

Cotton, S., G. Ford, S. Ford, and T. Hale. 2012. Internet use and depression among older adults. *Computers in human behavior* 28 (2): 496–499.

Dmitriev, M.E. 2005. Novyj strategicheskij vybor: mnimye i real'nye ugrozy dlya pensionnoj sistemy. In *Laboratoriya pensionnoj reformy. Informacionno-analiticheskij portal* http://pensionreform.ru/42409. Accessed 6 Dec 2017.

Grigoryeva, I.A., A.S. Bikkulov, and G.M. Cinchenko. 2014a. Starenie, mezhpokolennye vzaimodejstviya i zanyatost lyudej pozhilogo vozrasta. *Upravlencheskoe konsultirovanie* 12 (72): 101–110.

———. 2014b. Na puti k normativnoj modeli otnoshenij obshchestva s pozhilymi lyudmi. *Zhurnal sociologii i social'noj antropologii* 3: 151–167.

Information Web-Portal for the Aged People *Golden Age*. http://zolotoy-vozrast.ru/ Accessed 26 Nov 2016.

Internet polezen dlya pozhilyh lyudej. 2014. *Medicinskij onlajn zhurnal*, 14.08.2014. http://icj.ru/original-researches/6088-internet-polezen-dlya-pozhilyh-lyudey.html. Accessed 6 Dec 2017.

Kolesov, V. 2009–2016. *Virtualnaya shkola pensionera – buduschego i nastoyaschego*. http://www.ypensioner.ru. Accessed 6 Dec 2017.

Kroshilin, S.V., and E.I. Medvedeva. 2011. Vliyanie informacionno- kommunikacionnyh tekhnologij na formirovanie chelovecheskogo kapitala ili perspektivy postroeniya informacionnogo obshchestva v Rossijskoj Federacii. *Nacional'nye interesy: prioritety i bezopasnost* 41 (134): 22–30.

Laconi, S., N. Tricard, and H. Chabrol. 2015. Differences between specific and generalized problematic internet uses according to gender, age, time spent online and psychopathological symptoms. *Computers in Human Behavior* 48: 236–244.

Lysenko, E.A., and S.V. Fedoseeva. 2014. Na puti stanovleniya informacionnogo obshchestva: likvidaciya cifrovogo neravenstva sredi grazhdan starshego pokoleniya. *Informacionnoe obshchestvo* 1: 11–16.

Maleva, T., and V. Mau. 2013. *Chetyre Strategii Obespecheniya Starosti*. http://www.vladimirmau.ru/ru/rss/item/readarticles/tatyana_maleva_vladimir_mau_chetyre_strategii_obespecheniya. Accessed 6 Dec 2017.

Maleva, T. M., and O. V. Sinyavskaya. 2008. Nuzhno li povyshat zanyatost pensionerov? *DemoskopWeekly* 341–342. http://demoscope.ru/weekly/2008/0341/tema04.php. Accessed 6 Dec 2017.

Nevinnaya, I. 2015. Bolshinstvu naseleniya ne nravyatsya pensionnye ogranicheniya. *Rossijskaya gazeta* (79): 6650.

Obi, T., D. Ishmatova, and N. Iwasaki. 2013. Promoting ICT innovations for the ageing population in Japan. *International Journal of Medical Informatics*. 82 (4): 47–62.

Opros FOM: Pensionery rvutsya obratno na rynok truda. 2014. http://www.finmarket.ru/economics/article/3614224. Accessed 6 Dec 2017.

Opros FOM: O planah ne platit pensii pensioneram s zarplatoj vyshe 83 000 rublej. 2015. *Opros FOM*. http://fom.ru/Ekonomika/12228. Accessed 6 Dec 2017.

Opros FOM: Stazh dlya polucheniya pensii. 2012. Opros FOM. http://fin.fom.ru/ekonomika/10682. Accessed 6 Dec 2017.

Panina, T.S., and N.V. Pavelieva. 2014. Ispolzovanie informacionno- kommunikacionnyh tekhnologij v nepreryvnom obuchenii lyudej "Tretiego vozrasta". *Professionalnoe obrazovanie v Rossii i za rubezhom* 3 (15): 50–54.

Peterburgskaya pensionerka. *Informatsionno-poznavatelny zhurnal* http://pensionerka.spb.ru Accessed 6 Dec 2017.

Pokolenie-X. http://pokolenie-x.com. Accessed 6 Dec 2017.

Ryabova V. 2015. Gadzhety ne dayut pensioneram degradirovat – issledovanie. URL: http://d-russia.ru/gadzhety-ne-dayut-pensioneram-degradirovat-issledovanie.html Accessed 6 Dec 2017.

Troisi, J. 2007. *AGEING-A challenge and an opportunity for the countries of Eastern Europe, Caucasus and Central Asia.* Chisinau.

Universitet tretiego vozrasta. 2012–2017. http://u3a.ifmo.ru/. Accessed 6 Dec 2017.

V poiskah novogo «serebryanogo veka» v Rossii: faktory i posledstviya stareniya naseleniya. 2015. *Vsemirnyj bank. Obzornyj doklad.* http://www-wds.worldbank.org/external/default/WDSContentServer/WDSP/IB/2015/09/15/090224b0830dc63c/1_0/Rendered/PDF/Searching0for00ng000overview0report.pdf. Accessed 6 Dec 2017.

Vacek, P., and K. Rybenska. 2015. Research of interests in ICT education among seniors. *Procdings on Social and Behavioral Sciences* 171: 1038–1040.

Vse goda. *Socialno-consultativny resurs dlya lyudei starshego pokoleniya.* http://vsegoda.ru/ Accessed 6 Dec 2017.

Yashina, A.V. 2014. Informacionnye tekhnologii i transformacii sistemy obespecheniya bezopasnosti. *Voprosy bezopasnosti* 4: 104–130.

Chapter 9
The Emotional Experience of Old Age as a Result of Media Work

Age inequality in human communities, both in previous eras and today, are embodied in full measure by experience stratification – affects, feelings, and desires are proportioned "in the hands" of representatives from different socially-supported age groups. It could be stated that the age transition chart is paired with mastering new emotions and parting with several of those, which the new given age group does not welcome. It appears that this affirmation contradicts impressions about innate psychological personalitytypes (for example, temperament types), which form the nature of individual concerns and relations to the world. However, talking about age stratification of emotions, especially love in old age – as in general, about the sociality of emotions – suggests focus on an individual's fundamental dependency on cultural conventions, trickling in, according to the laws of living in a collective, extending to biological impulses.

How emotional experience programming takes place in an inequality opportunity society is a relevant research question for sociologists answering the call of affective turn and parting with the common perception of emotions as an individual's inherent universal answers to external stimuli.

9.1 Studying Emotional Inequality in Social Theory

Emotional inequality and exclusion is problematized within the stratified theory what is traditional for sociology, but innovative for the emotion studies. In Western sociology, several authors' works develop stratified emotional understanding, for example J. Barbalet (Barbalet 1998), R. Collins (Collins and Stephen 1990), and J. Turner (Turner 2010). Russian sociology also examines this topic: significant "groundwork" has been laid forth by the publications of O.A. Simonova (Simonova 2014, 2013, 2012, 2009), V. G. Nikolaev, creating an abstract on J. Barbalet's book,

© Springer Nature Switzerland AG 2019
I. Grigoryeva et al., *Elderly Population in Modern Russia*,
https://doi.org/10.1007/978-3-319-96619-9_9

"Emotions, Social Theory, and Social Structure: A Macro-sociological Approach" (Nikolaev 2002), N. S. Rosov (Rosov 2011) using for explanation the social dynamics of the concept of R. Collins "emotional energy."

Expression of emotion as a result of social learning is discussed today using the concepts of the sociology of culture. Culture has fundamental significance for our understanding of what constitutes emotions. Social feelings exist in inseparable connection with culture and are determined as "socially-constructed feelings templates, expressive gestures, and cultural ideas, organized around interactions with a social subject, generally another person" (Gordon 1981: 566). Social feelings, thus, are a social order phenomenon, since they are determined by culture, and depend on socialization and training.

Culture supplies people with a dictionary to describe our emotions, scenarios for expressing emotions, emotional norms, sanctions, and methods of regulating emotions. In this context we can say about emotional culture. The way that society's emotional culture is nowhere clearly fixed and dynamic makes study it a difficult task. Considering the importance of understanding emotional culture as the basis for participation in daily interactions, emotional socialization has decisive meaning in establishing emotionally competent subjects. Emotional socialization is the process through which individuals pass getting to know the culture of emotions. To reach emotional competency, children, adult and elderly people should socialize within the limits of a given society's emotional culture.

Emotional culture outlines what and how individuals can feel, occupying different statuses as a subject, as well as the emotional object/target. Socialization cohesion relates to how society's emotional culture determines emotions appropriate for people of various ages. Emotional competency expectations vary depending on age and socialization cohesion, determining the waiting periods for every emotion's competency.

For the topic "the emotional experince of old age" it is very interesting to review N.Denzin's ideas studied movies and emotions, believing that emotional socialization through institutions creates culture. These institutions are "groups or facilities clearly oriented on creating cultural ideas". Although N. Denzin focused on cinematography, there are other institutions as well. These, first and foremost, are mass media, the education system, and church, for example (Denzin 1990). He wrote that emotional practices demonstrated in film (for example, the gender aspects of behavior) encompass intimacy representations. Films provide special images of romantic relationships. These examples are integrated into society's emotional culture and impact how contemporaries view love and intimacy. Thus, studying emotionality should always be connected to analyzing cultural, historical, and institutional context.

Emotions and feelings are social, meaning that they cannot be reduced to a particular person's physical manifestations. In study, cultural limitations are needed as well as how people, orienting on them and directing their emotions leads to feelings corresponding to social expectations. Our emotional experience reflects socially defined rules, like how we should feel in a particular situation.

9.2 Movies as Research Material

Despite the fact that films are most frequently studied as works of art, they are also social artifacts creating an ideal environment for broadcasting symbols and images, behavior strategies, and ideas. Movies from different decades are irreplaceable as historical documents, giving valuable visual and auditory information, for example about the rules of romantic relationships in each socio-historical period.

Movies are a part of daily life, expressing the diversity of human existence. But film is not only a preferred medium for expressing public imagination – wishes, dreams, and fears, i.e. the collective subconscious in which one can find trans-historical Jungian archetypes – it is also a channel for manufacturing meanings that legitimize ways of seeing different objects. By this direction of gaze, allowing some to observe and others to be observed, social researchers have criticized film.

Consistent criticism of the media, particularly film, as representational instruments affixing the hegemony of men, was feminist theorists' argument basis in the 1970s–1980s. Many ideas about the power of cine-optics, establishing inequality and excluding certain groups to benefit others, resounded 30 years ago in relation to gender, are relevant today to understand different inequalities, for example, ageism and the hegemony of the young.

Thanks to work in the 1970s, we determined representation as a process through which culture subjects use language (any system of symbols) to create meaning. Representational objects do not have meaning by themselves: it is born in the process of interpretation and communication, coding and decoding texts, and depends on cultural context (Usmanova 2001: 449). Consequently, film representations of someone or something are not the image of phenomena, but "more likely the active process of selection and representation, structuring and formation, the process of vesting something with meaning" (Usmanova 2001: 452). Representations create gender, ethnic, national, and age categories, thanks to which borders are supported and relationship order affirmed.

In light of ideas on the representational work of media, social researchers began to focus attention not so much on the specifics of discriminated groups' images, as on processes of creating and using these images. One can take to arms the following critical feminist theorist logic: "gender analysis of representations genuinely started when instead of "female images" researchers turned to studying "women as an image." Interest in women's positions in the narrative and in separate genres, attention to ways through which patriarchal society not only structures content, but determines the way of seeing, the accent on images "constructedness" and "generatedness" – all this enabled forming deconstructivist analysis of mass media texts" (Usmanova 2001: 451).

9.3   Research Method

Working with movie materials, I relied on structural text analysis, approved by V. Y. Propp (2001). V. Y. Propp studied fairy tales and proposed segmenting their texts into (1) actions or functions of characters (and sequence of functions), (2) subjects of actions or characters and their attributes; (3) supporting elements existing to interconnect functions. This approach reveals variable and invariant elements of a narrative and their connections, based on which it is possible to interpret the work's intension and understand the cultural meaning.

We apply Propp's method chiefly to analyze mass cinema, the plot of which is intended to be accepted by the average viewer. Mass cinema, like the folk tale, is targeted at an audience defined as "all viewers."

The subject of my research were Russian films (not television shows) that have a plotline with romantic relationships between people of ages frequently denoted by the chronological definition of old age as "pensionary." Setting the task of tracing the dynamics of representing romantic experience of heroes at this age, I used Soviet and post-Soviet film materials, differing by genre, filmed starting in the 1970s. This decade was not selected accidentally as a milestone.

The standards of love and sexuality accepted today began to emerge specifically in that period. As I. S. Kon wrote, "Beginning in the early 1960s, as soon as the repressive regime weakened, sexual discourse began to grow. In addition, it was revealed not only the country's monstrous underdevelopment, but also that despite all repression and social isolation from the Western countries the main trends of dynamic changes of sexual behavior in the USSR were the same as amongst our ideological enemies: lowered sexual debut age limit, emancipation, separating sexual motivations from matrimonial, increased number of divorces, premarital and extramarital conceptions and births, increased interest in erotica, and women's resexualization, etc. Movements in this direction began not in the Perestroika and Glasnost era, but already in the 1960s and especially the 1970s" (Kon 2004: 65).

Dynamization of love/sexual fields in the post-Soviet period is inseparably linked to media processes. N. N. Kozlova writes about this, analyzing the everyday life in this era: "Departure from Soviet identity occurs in different ways. It is impossible to diminish the meaning of the arrival of new visual means of communication, which instead of the certainty promised by Bolshevist enlightenment, offered a collage of lifestyles and circumstances" (Kozlova 1998: 169). Perceptibly, in the 1970s, several collective attitude reference points inherent to the "classic Soviet era" were lost and new, often more liberal ones were found, which film allows us to see.

Film selection included 16 works: from those filmed in 1971, to those coming to theater in 2005 (see the list below). In addition, the place of romantic lines in relation to film's plot could be major (for example, the 1984 film "Let the Charms Last Long," directed by Y. Lapshin or the 2000 film "The Garden was Full of Moon," directed by V. Melnikov), as well as secondary (for example, the 1981 film "Family Relations" directed by N. Mikhalkov, or the 1991 film "Promised Heaven," directed by E. Ryazanov).

9.4   Love 60+ on Screen: Research Results and Their Discussion

9.4.1   Vectors of Developing Romantic Relationships

The systematization of heroes' actions (or functions, per V. Propp) gives two main ways of onscreen love line development in older age:

1. The all-age way of relationship development
   - Acquaintance
   - Flirtation
   - Relationship Crisis
     • Favorable Crisis Resolution (Relationship Continuation)
     • Unfavorable Crisis Resolution (Separation)

2. The specific way of developing relationships
   - Life-long Infatuation
   - Optionally – Running into an Old Flame
   - Relationship Crisis
     • Favorable Crisis Resolution
     • Unfavorable Crisis Resolution

Thus, Russian cinema represents two ways of developing romantic relationships in old age. The first way is universal for all ages, transitioning from acquaintance to "the crystallization of love" (a Stendhal expression denoting ascribing the best qualities to the object of love) (Vilyunas and Gippenreiter 1984), later to experiencing quarrels and misunderstandings, and finally, possibly, favorably continuing relations or heroes' separation.

The second romantic experience option can be called "specific," since it is connected with preserving the feelings of heroes over the duration of their entire adult life without union. This emotional experience is characterized by sad experiences, feelings of loss, which are aggravated by the lag between first meeting and today's day of the heroes. Representation of missed opportunities, of past irreversible meetings create a stable image of old age as time "when everything has already happened." I suggest that this love line scenario in film – despite all the film's artistic merits – blocks confidence for positive feelings in the third age. This block is reminiscent of the "willingness to resign," characteristic of Russian elderly.

Reflecting on the structural functions of films with romantic storylines directs the researcher to illuminate the question of sexuality manifestations for heroes from the older age groups. In truth, films about love are always connected with showing different pictures of characters' closeness. This is not necessarily physical intimacy, but is a demonstration of sensual tactility – embraces and kisses, first of all. Russian cinema about older lovers, both Soviet and modern, eliminates displays of third age sexuality. We see the action of strict cultural standard, blocking representation of aging partners' physical attractiveness and sexual healing.

Today, despite widespread openness in relation to sexual problems, the perspective remains that older people displaying sexuality is something not normal, stupid, or awkward. This occurs because society thinks of older people as asexual. In Russian culture, the aging body is stripped of gender signs, and it is assumed that attractiveness and erotic attributes of physicality disappear with age. Repression of aging women was and remains especially severe, still connected with insistent recommendations to fulfill, at last, the housekeeping position "befitting their years." The positive aspect of this process has been the turn to studying third-age women specifically, i.e. the "sandwich generation" (a term introduced into circulation by D. Miller in 1981), wedged in between duties related to children and grandchildren, as well as to fourth-age parents (Morozova 2014; Zdravomyslova and others 2010).

This relationship rejects older people's sexual lives, and social representations in the form of popular jokes, superstitions, and taboos encourage exclusion from the erotic sphere. It has long been noted that comical reduction and the demonization of older women traditionally coexist in culture (Bocharova 2000). And if the "old witch" is an unambiguously ageist image, unacceptable in modern tolerant society, then the "infatuated old woman" is an image bordering on absurd....

S. Gonzalez (Gonzalez 2007) applied age classification aimed at unifying the life cycles of sexuality. His goal was to show how the meanings of age in culture are determined by the dynamics of sexuality. Our bodies might be desexualized (in the cases of older people), sexual (in youth), asexual (as with children), sexually legitimate (as in marriage), and sexually "destitute" (when our body decays). Making classification demonstrates that understanding age and sexuality are closely interwoven in our daily life.

Nevertheless, it is impossible not to talk about several movements blurring the taboos of post-Soviet cinematography: the veil over forbidden topics are raised by comedic films (for example, the 1999 movie "Quadrille," directed by V. Titov). In toe with humorous modality, the topic of sexuality in older age resounds. This trend is entirely explainable, as joking forms help decrease the intensity of taboo and avoid sanctions for violating rights.

9.4.2   Heroes of Romantic Relationships

The key characters of films with a romantic plot, focusing on the feelings of older people are: the hero, the heroine, the adversary (maybe a deceased spouse), and grown-up children and grandchildren. Talking about the heroes of Russian films experiencing love in the third age, it is worth mentioning their typical attribution as the so called "younger elderly." Meaning that the characters' ages shift toward the border of "a little past 60," and older age groups are excluded as subjects of Cupid's arrow. Although work with the topic of romantic feelings of "grown-up" older groups (those whose age is closer to 80) would an interesting creative task for a screenwriter, director, and actors, not to mention the social significance of this topic.

One of the rare attempts to talk about the feelings of "very grown-up" heroes can be considered the 1983 film "I zhizn, i slyozy, i lyubov" from director N. Gubenko.

The emotional experience borders of the oldest age groups represent a complicated research and ethical topic, while if we are talking about old age as a cinematic image, then it is aesthetic as well. It would be worth acknowledging that in all historic periods aged people have been distinguished from the remaining population, first due to their relatively small quantity (children and youths were predominant as an effect of high birthrate), second their appearance, and third their experience and worldly, even prophetic, wisdom.

True, in evaluating older people such characteristics as "senility" and "wisdom" have always competed. In the long history of culture there is no united basis (archetype) for aging. In social consciousness, from ancient times through today, one can distinguish two contradictory images of old age: the image of the wise elder and the image of the decrepit geezer. These two representations have successfully competed for many years running. They have spawned numerous myths of old age, and consequently, many social problems are believed to be connected to it.

It is difficult to say from which historical period the elderly became personified as "others," and often with the specific view point of physical weakness and decrepitude. Visibly, development of pension insurance at the end of the nineteenth century played a deciding role in this. The situation began to substantially change in the second half of the twentieth century when, due to decreased childhood mortality, the average life expectancy increased. The number of older people in society became noticeably larger and along with aging appearance became the object of "soft" (cosmetic) and "hard" (surgical) correction, while the worldly wisdom stockpiled over the years began to quickly lose meaning in the time crunch of rapid social changes.

Probably, this is the reason we observe changes today in normative intergenerational relationships principles, which in the observable historical past issued from the principle that "old age is the time for departing from social activeness." Old age itself is shifted farther and farther back along with pension age – for now, just in the West, but soon in Russia as well. There are increasingly more cheerful, healthy people over the age of 70, who still do not wish to retire....

These changes require understanding the fact that people of any age can be socially and emotionally active, and choose forms of this activeness for themselves. In light of our interest in film, it is also obvious that showing older people's heart's desires breaks the negative cultural fixation on the "otherness" of old age and, subsequently, should decrease fear of this age as a period of negative corporeal transformations and gradual social isolation and departure.

We see emotional normalcy "habitualization" and showing variety amongst elderly people as an important task for cinema.

Analyzing film characters also lets us talk about the characteristics of "love triangles" in plots, the role of children and grandchildren, the rules of showing initiative in developing relationships, and the traditionalness of heroes' sexual orientation. "Triangles" as a kind of social rivalry are incorporated in films about the third age, however their specifics might comprise competition not with a person, but with the memory of a deceased spouse (for example, the films "Fathers and

Grandfathers" from 1982 and "Winter Romance" from 2004). Thus, the topic of death and loss intrudes into the space of romance, again programming the emotional experience of old age.

The standard plotline includes children and grandchildren in the dynamics of grandmothers' and grandfathers' romantic relationships. Intergenerational interactions onscreen can be indicative of ageism, due to the fact that the younger generation either supports or criticizes and judges their older relative's feelings.

Using V. Propp's terminology, one can say that in narratives about older aged heroes' love, their children act as antagonists, presenting obstacles, and leading the romantic heroes through trials. Normalcy/abnormality of romantic emotions in the third age become the object of evaluating, first of all, not peers but other generations, which also reinforces the image of being "dependents."

According to the conclusions of several specialists-gerontologists on the meaning of family, reorienting on internal family relationships in later years is a natural stage in of an older person's mental life. Saratov sociologists write that family values in this period are of primary significance. The psychologist M. Ermolaeva evaluates immersion in family as a beneficial fact and demonstrates the level of this immersion. Based on her research, statements made by people of the older generation testify to the fact that they are engaged in their loved ones' problems, perceive these problems as their own, and often correlate life goals and plans with events occurring in the younger generation's life (Ermolaeva 2002).

The renowned book "Labyrinths of Loneliness" (Pokrovsky 1989) also talks a lot about the "empty nest" phenomenon and other variations of experiencing loneliness, primarily, by elderly women. But first, what is loneliness – solitary living or a multigenerational family forced to live together with relatives? Understandably, there are dark nuances here, and sociologists' attempts to calculate where and how many older people are alone creates the impression of conclusive reductionism.

Aside from this, construing loneliness as a problem is the reverse side of the "sandwich generation" problem, which we have mentioned several times earlier. Which burden is heavier – family day-to-day care, serious emotional labor (which, according to A. Hochschild, launched this discussion), or the "void of loneliness?" And although "the everyday is elusive for us, we are inclined to believe that it is simply life, the natural norm of practical existence" (Lotman 1994: 10), but transforming the everyday and life, the opportunity of dedicating them to "caring for oneself" also requires justification. Yes, and loneliness can now be called "living solo" and is considered to provide a multitude of new opportunities (Klyajnenberg 2014).

Thus, as we see from the representations in films with romantic plots, in order to have the right to feelings, the older generation needs a certain amount of emancipation. Or there should be more films about the fact that "all ages are under love's power."

9.4.3   Retirement as a Key Event Structuring Emotional Experience in the Second Half of Life

Corresponding with the V. Propp's narrative analysis logic, along with the functions on which's fulfilment the plots' development depends, there are supporting elements serving to connect heroes' fateful actions. Supportive elements in the analyzed films could be considered retiring, moving to a retirement home, or taking a trip to a health spa. It is these particular events that initiate the aged heroes' meetings, acquaintance, and devotion (for example, "Granddads-Robbers" from 1971, "Prodlis, prodlis, ocharovanie" from 1984, and "December 32nd" from 2004).

We see that films in which aging characters participate inevitably editorialize retirement. As a matter of fact, retirement has long been considered the first sign of aging, as we have already mentioned. So, one can assert that receiving the status of retiree, meaning that a person is not working, is a necessary component of the old age image in our current collective representations. Calling someone a retiree unambiguously involves our quantity of years, and not, for example, disease, although there are disability pensions.

Applying neutralizing chronological definitions of old age, connected with the retirement system, leads to the recent undesirable social treatment (Shapiro 1980: 10). In public discourse, members of older age groups appear to be those who are much weaker, and require special security, from the one hand, and those who do not need much in life any longer, from the other. Elderly romantic relationships and love collisions are built against this "retirement" background, regulating emotional needs.

E. Durkheim defended the idea of social phenomena's irreducibility on a person's individual qualities, determined by his mental nature. The affective turn in sociology modified the Durkheimian premise, turning it into an assertion about social organization forming and displaying human feelings. This signifies that certain spiritual impulses and emotional bursts are cultivated, while others are suppressed by the collectives in which an individual lives and acts, and social status provides not only rights and imposes duties on action, but also formats experience alongside this.

The rapid development of informational and communicative technologies led to different media being limitlessly distributed and more accessible to audiences than in previous eras. One of the characteristics of this audio-visual content is distributing cultural scenarios for displaying emotions.

In particular, cinematic images of love, sexuality, and physical attractiveness are popular in mass cinematography, and can be considered a factor programming emotional experience, furthermore, representatives of different age groups. In modern times, studies on the influence of media representations of romantic emotions are implemented amongst teenagers, while third age heroes and their audience remain essentially under the radar.

Russian films, both Soviet and post-Soviet, are important emotional socialization agents. They possess the power to determine which emotions are permissible and correct for people of different age groups. The appearance and manifestation of romantic feelings by people who have transitioned past the border of old age contradict widespread current social expectations, which often have an openly repressive nature.

Sexuality is most strictly controlled for heroes "of a certain age." In particular, erotica in Soviet culture was always understood as the erotic desire of young people, cinematic culture did not offer chances for eroticism to those who had crossed the threshold of youth. This associates older people's sexuality with the idea of the illegality of sex and reinforces images that spawn feelings of loneliness, loss, and even depression.

Sociological analysis allows for considering a person's life as a path, which includes needs and expectations in relation to age, including age-relevant emotions. "You cannot do (blank) until age X," and "at age X you should (blank)," or "acting your age" – these are all examples of how society understands life cycle with its limits and roles. Into the understanding of age enter cultural emotional scenarios, broadcasted via the most sensitive medium – the cinema.

9.5   Conclusion

Analyzing film characters lets us also talk about the characteristics of "love triangles" in plots, the role of children and grandchildren, the rules of showing initiative in developing relationships, and the traditionalness of heroes' sexual orientation. "Triangles" as a kind of social rivalry are included in film about the third age, the specifics of which, however, might be competition not with a person, but with the memory of a deceased spouse (for example, the films "Fathers and Grandfathers" from 1982 and "Winter Romance" from 2004). Thus, the topic of death and loss intrudes into the space of romance, again programming the emotional experience of old age.

The standard plot course includes children and grandchildren in the dynamics of romantic relationships of grandmothers and grandfathers. Intergenerational interactions onscreen can be indicative of ageism, due to the fact that the younger generation either supports or criticizes and judges the feelings of their older relatives. Normalcy/anormality of romantic emotions in the third age become the object of evaluating, first of all, not peers but other generations, which also affixes the image of dependents.

Physical age changes present a threat for those who are young and prompt low self-esteem amongst older people in relation to their corporeal characteristics. Gender roles of elderly film heroes are connected, alongside previously mentioned ideas, with the traditional theatrical and cinematic role of comical grannies and geezers. The comedic mas of the aging hero does not lose relevance in the modern cinema, which also establishes a "lowered" cultural image of old age and testifies to several important bans on feelings.

Following this dramaturgic logic, we are predisposed in life to play the roles of romantic heroes and then comedic geezers and grannies. Exaggerated admiration of youth in Russian culture creates a bizarre and contradictory cultural background of a fragmented, divided society which is constantly contradicting itself and cannot find development ideas.

Analyzed films:

1. Kill the Carp, N. Ardashnikov (2005)
2. Winter Romance, N. Rodionova (2004)
3. December 32nd, A. Muratov (2004)
4. Moscow Elegy, V. Akhadov (2002)
5. The Garden was Full of Moon, V. Melnikov (2000)
6. Old Hags, E. Ryazanov (2000)
7. Quadrille, V. Titov (1999)
8. Russian Love, E. Matveev (1995)
9. Promised Heaven, E. Ryazanov (1991)
10. I Still Love, I Still Hope, N. Lyrchikov (1984)
11. Let the Charms Last Long, Y. Lapshin (1984)
12. Life, Tears, and Love, N. Gubenko (1983)
13. Fathers and Grandfathers, Y. Egorov (1982)
14. Family Relations, N. Mikhalkov (1981)
15. For Family Reasons, A. Korenev (1978)
16. Granddads-Robbers, E. Ryazanov (1971)

References

Barbalet, J.M. 1998. *Emotion, social theory, and social structure: A macrosociological approach.* Cambridge University Press.

Bocharov, V.V. 2000. *Antropologiya vozrasta.* St. Petersburg: Izd-vo S.-Peterb. un-ta.

Collins, N., and R. Stephen. 1990. Adult attachment, working models, and relationship quality in dating couples. *Journal of Personality and Social Psychology* 58 (4): 644–663.

Denzin, N. 1990. Understanding emotion: The interpretive-cultural agenda. In *Research agendas in the sociology of emotions*, ed. T.D. Kemper, 85–116. Albany: State University of New York Press.

Ermolaeva, M. 2002. *Prakticheskaya psihologiya starosti.* Moscow: Izd-vo EKSMO Press.

Gonzalez, C. 2007. Age-graded sexualities: the struggles of our ageing body. *Sexuality and Culture* 11: 31–47.

Gordon, S.L. 1981. The sociology of sentiments and emotion. In *Social Psychology: Sociological Perspectives*, ed. M. Rosenberg and R.H. Turner, 562–592. New York: Basic Books.

Klyajnenberg, E. 2014. *Zhizn solo. Novaya socialnaya realnost.* (Trans: Andreev, A.) Moscow: Alpina.

Kon, I.S. 2004. *Seksualnost i kultura.* St. Petersburg: SPbGUP.

Kozlova, T.Z. 1998. *Sotsialno-istoricheskaya antropologiya: Uchebnik.* Moscow: Klyuch-S.

Lotman, Y.M. 1994. *Byt i tradicii russkogo dvoryanstva (XVIII - nachalo XIX veka).* St. Petersburg: Iskusstvo-SPb.

Morozova, T. 2014. Sendvich-pokolenie ne vybiraet. *Moj rajon* (23.06.2014).

Nikolaev, V.G. 2002. Referat knigi Dzh. Barbaleta "Emocii, socialnaja teorija i socialnaja struktura: makrosociologicheskij podhod". *Sotsiologicheskoe obozrenie* 2 (9): 3–9.

Pokrovsky, N.E., ed. 1989. *Labirinty odinochestva*. Moscow: Progress ISBN: 5-01-001589-7.

Propp, V.Ya. 2001. *Morfologiya volshebnoy skazki*. Moscow: Izdatelstvo Labirint.

Rosov, N.S. 2011. Emocionalnaja energiya: istoriko-sociologicheskiy analiz. *Sotsiologicheskiye issledovaniya* 2: 12–23.

Shapiro, V.D. 1980. *Chelovek na pensii: sotsialniye problem i obraz zhizni*. Moscow: Mysl.

Simonova, O.A. 2009. Sociologicheskoe issledovanie emocij v sovremennoj amerikanskoj sociologii: konceptualnye voprosy. *Sotsiologichesky Yezhegodnik.* 1 (1): 199–225.

———. 2012. Kontseptsiya emotsionalnogo truda Arli R. Hohshild. Antropologiya professiy: granitsy zanyatosti v epohu nestabilnosti. Moscow: OOO Variant: 75–96. http://www.hse.ru/pubs/share/direct/document/77455935

———. 2013. Emocionalnyj trud v sovremennom obshchestve: nauchnye diskussii i dalnejshaya konceptualizaciya idej A.R. Hohshild. *Zhurnal issledovaniy sotsialnoy politiki* 11 (3): 339–354.

———. 2014. Styd i bednost: posledstvija dlya socialnoj politiki. *Zhurnal issledovaniy sotsialnoy politiki* 12 (4): 539–554.

Turner, D. 2010. Qualitative interview design: A practical guide for novice investigators. *The Qualitative Report.* 15 (3).

Usmanova, A. 2001. In *Gendernaya problematika v paradigme kulturnyh issledovaniy. Vvedeniye v gendernye issledovaniya*, ed. I.A. Zherebkina, vol. 1. St. Petersburg: Aletejya.

Vilyunas, V.K., and Yu.B Gippenreiter, eds. 1984. *Psychology of Emotions*. Moscow: Izd-vo Mosk. un-ta.

Zdravomyslova, E., V. Pasynkova, A. Temkina, and O. Tkach. 2010. *Praktiki i identichnosti. Gendernoe ustrojstvo*. St. Petersburg: Izd-vo Evropejskogo universiteta v SPb.

Chapter 10
Conclusion

Current discourse about the elderly in modern science, undoubtedly, creates a contradictory, tense situation for elderly people, other social age groups, and for society as a whole. Of course, the question about what can and, obviously, should elderly people do for themselves is imperative to seriously discuss.

We must, because duty in relation to oneself and one's family remains key for each normal person, the debt to remain a subject of his or her own life. In the context of macrosocial movements, the appearance of studies seems relevant which problematize the immutability of age boundaries and statuses, rejecting perception of the aging population process as a global problem and risk.

Following awareness that class and property are insufficient for explaining contemporary social-economic stratification, gender actualization, and ethnicity as notable inequality parameters, concentration is growing on age and social-age classes/groups/generations (Bollinger and Malek 1989: 116–121).

A person retiring today is not considered economically productive and is thus devalued. We interpret this as modern mass culture's retreat from the principle of anthropocentrism in favor of socio-centricity and labor-centricity, as a result of which a person began to offer value only during the period of active labor activity. To justify "expenses on retirees," instead of explaining that retirement is something they earned, the thesis is used that they worked for the good of society. By him or herself a person is insignificant in socio-centric conversation.

This redaction of modernity as a consumer society changes little for those who cannot be active users. If a person cannot be an active customer, then he or she can be an active social services consumer. The role of elderly people in the economy is connected not just with extended employment, but with developing their demand and consumer practices for medicine, as well as pharmacology, cosmetology, tourism, and other service infrastructures. Activating the older population is often associated specifically with increased leisure consumption demand.

We came to the conclusion that computer technology use is becoming a rapidly widespread practice, confirming an older person's "modern-ness." Training for such skills is entirely accessible and is even becoming trendy amongst older people.

© Springer Nature Switzerland AG 2019
I. Grigoryeva et al., *Elderly Population in Modern Russia*,
https://doi.org/10.1007/978-3-319-96619-9_10

Visibly, it corresponds to the humanistic premise of the eternal education concept, oriented on participants' interests rather than on functional use.

However, ICT practices used by the elderly are for now weakly oriented on cyber-aid and self-sufficiency. Researchers note that "cyber-assistance for socially vulnerable groups is a mechanism of social policy not free from contradictions;" however, potentially, it has a great positive effect, as foreign researchers note (Pleace et al. 2003).

Elderly people's problems, on the one hand, are specific to that social age group. On the other hand, they are very similar to other social groups' problems. Difficulty accessing professional work, especially that which is well paid, also affects younger people; problems with health and medical aid are no less relevant for children, as well as for other age groups; changes in the family and habitual sphere and low life quality level also often "do not have an age."

In addition, all these problems bear a deeply Russian nature. Many of the world's experience similar complications countries to one extent or another and conduct active social policy to resolve them. However, the older generation in developed countries has already long been an example for Russia to follow. Perceptibly, the reason for this is encompassed not just in the economic sphere, but principally in the cultural and spiritual sphere.

Knowledge about older people's real situations are fairly contradictory in modern times, but are united by general acceptance of the segregating, condescending view of elderly problems "from the top down." Many, even scientific works, include ambivalent stereotypes that are unyielding and widely-distributed in Russian society and clichés, typifying schemes on questions of old age and aging, the history of which began in ancient traditional societies.

But "for theories developed in the frameworks of different paradigm conditions, no procedure exists to help unambiguously determine which of them is genuine and which is false. The very solution of such a problem begins only after one of the paradigms is chosen. But this choice is a volitional act and, from the point of view of classical rationality, is irrational.

Here one can talk about the rationality of the non-classical kind – the rationality of choosing a model of community's "living world". Within the limits of this rationality it is impossible to prove which of the constructed paradigm systems is true, however as is valid we can only choose one of them (but not both together), since without this choice it is impossible to determine the real order of social life.

The will to adopt certain ontological postulates is stipulated by the (conscious and unconscious) ambition for social consolidation, to reach agreement on the most important life questions, from which only a small part belong to the authority of proper science" (Lipsky 1999).

Shifting paradigm, we gradually begin to understand that older people and old age are our future, which in many ways depends on how we "play" or intend to play the role of elderly person, from which age we identify ourselves as elderly, and which expectations and fixtures we have along with our environment relating to this stage of life.

The authors wanted to show that social exclusion phenomena described in monographs, loneliness, and passivity among the elderly population are products of society, the result of practical activity, and mental design of everything without excluding social and age groups, including older people. Of course, our research was conducted in Saint Petersburg, which can be considered advanced in relation to other forms of work with the elderly and, obviously, the situation in many places is worse. Instead, Saint Petersburg can be considered a model, as an experiment sent up by the government, the positive and negative instances of which are constructive.

Modern aging society faces the necessity of replacing behavior models in new socio-historical conditions. The strategy of elderly displacement, service, and incomplete participation in society life, is essentially conflicting and unproductive – it is a blind strategy, reinforcing modern ineffective social constructs of old age.

The strategy of cooperation, acceptance, and solidarity, which recognizes multilinearity, continuity, selectivity, pluralism of individual development due to the subject's unique activity, and the influence of the environment (the general socio-historical and cultural background of all generations), represents the only true step toward a future sustainable society.

Schemes of age periodization, in the frameworks of which we "customize" life for the average individual, cannot predominate over free self-definition growing from population to population. Elderly people are the titular age group of the future. Many modern older people, especially in large cities, have time in retirement – this is a time of unprecedented freedom. After all, in addition to low current income, some have accumulated property, have rental income, and more free time than ever before.

This is important to remember when talking about elderly people's position in modern society. However, this is not a summons to return to a social guardianship mechanism, but the opportunity to newly view the elderly generation's constructive potential, which quite possibly could begin to build a "government of wisemen," the meritocracy about which ancient philosophers dreamed.

References

Bollinger, S., and B. Malek. 1989. Myshlenie mezhdu utopiej i real'nost'yu. Mirovozzrencheskie pozicii al'ternativnogo i ehkologicheskogo dvizheniya v FRG. In *Novye social'nye dvizheniya i sociokul'turnye eksperimenty*, vol. 1, 116–121. Moscow.

Lipskij, B. I. 1999. Edinstvo intelligibelnogo mira i mnozhestvennost «zhiznennyh mirov». *Vestnik Sankt-Peterburgskogo gosudarstvennogo universiteta*. Ser. 6. 1(No 6): 3–6.

Pleace, N., R. Burrows, B. Loader, T. Mullen, and S. Nettleton. 2003. From self-service welfare to virtual self-help? In *Information and communication technologies in the welfare services*, ed. E. Harlow and S. Webb, 183–198. London: Jessica Kingsley.

Index

© Springer Nature Switzerland AG 2019
I. Grigoryeva et al., *Elderly Population in Modern Russia*,
https://doi.org/10.1007/978-3-319-96619-9

Printed by Printforce, the Netherlands